GENETICS, THEOLOGY, *and* ETHICS

GENETICS, THEOLOGY, *and* ETHICS

An Interdisciplinary Conversation

Lisa Sowle Cahill, editor

A Herder & Herder Book
The Crossroad Publishing Company
New York

The Crossroad Publishing Company
481 Eighth Avenue, New York, NY 10001

This book is set in 10.5/13.5 Sabon.
The display type is Trajan and Futura.

Printed in the United States of America

Library of Congress Cataloging-in-Publication Data

Genetics, theology, and ethics : an interdisciplinary conversation / Lisa Sowle Cahill, editor.
p. cm.
Includes bibliographical references and index.
ISBN 0-8245-2269-9 (alk. paper)
1. Human genetics—Moral and ethical aspects. 2. Human genetics—Religious aspects. 3. Bioethics. I. Cahill, Lisa Sowle.
QH438.7G466 2005
174'.957—dc22

2005004019

1 2 3 4 5 6 7 8 9 10 09 08 07 06 05

Contents

Contributors

HASNA BEGUM, Ph.D. (Monash), specializing in ethics, was a professor of philosophy (now retired) at the University of Dhaka. Currently, she is one of the board members of the International Association of Bioethics (IAB). Among her publications is *Women in the Developing World: Thoughts and Ideals* (Apt Books, 1990). Dr. Begum is an advocate for basic health care for the poor of Bangladesh and a vocal critic of the economic and military forces that combine to exclude many from access to health care.

LISA SOWLE CAHILL is J. Donald Monan, S.J., Professor of Theology at Boston College, and director of graduate programs in theology. She is a past president of the Catholic Theological Society of America (1992–93), and the Society of Christian Ethics (1997–98). Among Lisa Cahill's recent works are *Bioethics and the Common Good* (Marquette University Press, 2004); *A New Bioethics: Theology, Justice, and Change* (Georgetown, projected for 2005). She is a member of the American Academy of Arts and Sciences, and of the March of Dimes Bioethics Advisory Committee. She serves on the editorial boards of several journals, including the *Journal of Religious Ethics*, the *Journal of Law and Religion*, and the *Kennedy Institute of Ethics Journal*.

GERRY EVERS-KIEBOOMS, Ph.D. in Psychology, is a professor in the Department of Human Genetics in the Faculty of Medicine in the Catholic University of Leuven, Belgium. She is the head of the Psychosocial Genetics Unit of the Center for Human Genetics in Leuven, Belgium, and member of the board of the Genetic Clinic. The Psychosocial Genetics Unit has three major objectives: research, counseling, and genetic education. She is a member of several scientific societies, including European Society of Human Genetics, the Belgian Society of Human Genetics, and the World Federation of Neurology Research Group on Huntington's Disease. She is a member of the Belgian Advisory Committee in Bio-ethics.

MÁRCIO FABRI DOS ANJOS is a past president of SOTER, the Brazilian Society of Theology and Sciences of Religion, and a past member of the Brazilian National Commission for Research Ethics. Currently he is serving as professor of theology at the University Assuncao, in San Paulo, Brazil, and director of the Alfonsianum Institute for Ethics and Theology. He is also a member of the Committee of Bioethics of the Regional Council of Medicine in São Paulo. He is the author of several books and articles.

KEVIN T. FITZGERALD, S.J., Ph.D., is the Dr. David Lauler Chair of Catholic Health Care Ethics in the Center for Clinical Bioethics at Georgetown University. He is also an associate professor in the department of oncology at the Georgetown University Medical Center. His scientific research efforts focus on the investigation of abnormal gene regulation in cancer and research on ethical issues in human genetics. He has published both scientific and ethical articles in peer-reviewed journals, books, and in the popular press. He does regular briefings for the United States Catholic Conference and for various congressional members and committees. He is a founding member of Do No Harm, a member of the ethics committee for the March of Dimes, and a member of the American Association for the Advancement of Science Program of Dialogue on Science, Ethics, and Religion.

BART HANSEN (b. 1977) is a research fellow of the Belgian National Foundation on Scientific Research (FWO-Vlaanderen) at the Center for Biomedical Ethics and Law, Faculty of Medicine. He is a member of the Faculty of Theology, K.U. Leuven, and is currently preparing a doctoral research project on the theology of human stem cell research.

JAMES KEENAN, S.J., was a professor at Weston Jesuit School of Theology from 1991 to 2005, when he became a professor of moral theology at Boston College. He has served on the editorial board of *Theological Studies* since 1991 and the board of directors of the Society of Christian Ethics (2001–2005). He is the editor of the series Moral Traditions at Georgetown University Press. *Moral Wisdom* and *The Works of Mercy* are Father Keenan's newest books. His earlier works include *Goodness and Rightness in Thomas Aquinas's Summa Theologiae*, *Virtues for Ordinary Christians*, *Commandments of Compassion*, and, more recently, with Daniel J. Harrington, S.J., *Jesus and Virtue Ethics*.

BARTHA MARIA KNOPPERS, Canada Research Chair in Law and Medicine, is a professor at the Faculté de Droit, Université de Montréal, senior researcher at the Centre for Public Law (C.R.D.P.), and is an *Officer* of the Order of Canada. Currently chair of the International Ethics Committee of the Human Genome Organization (HUGO), she is cofounder of the International Institute of Research in Ethics and Biomedicine (IIREB) and a codirector of the Quebec Network of Applied Genetic Medicine (RMGA). She was named to the Board of Genome Canada in 2000. In 2003, she became director of the international Public Population Project in Genomics (P^3G).

DIETMAR MIETH has been professor for theological ethics/social ethics of the Catholic Faculty of Theology at the University of Tübingen since 1981. From 1990 to 2001, he served as director of the Center "Ethics in the Sciences and the Humanities" at the University of Tübingen. He has served on various ethics advisory groups, including "Ethics in the Sciences and in the New Technologies" of the European Commission in Brussels (1994–2000) and the ethical advisory group of the Federal Ministry of Health (1998–2002). He has written twenty-seven books, edited fifty books, and is coeditor of the twelve-volume collection "Ethics in Sciences." His most recent book is *Mystik und Lebenskunst* (Düsseldorf: Patmos Verlag, 2004).

PAUL SCHOTSMANS is professor of medical ethics in the faculty of medicine of the Catholic University of Leuven, Belgium. He is a priest of the archdiocese of Mechelen-Brussels. He is director of the Center for Biomedical Ethics and Law (faculty of medicine, K.U. Leuven) and president of the Department of Public Health. He is also past-president of the European Association of Centers of Medical Ethics and an elected member of the Board of the International Association of Bioethics. He serves as a member of the Belgian Bioethics Advisory Committee. His main research topics are axiology and biomedical ethics. He coedited, with Reidar K. Lie, *Healthy Thoughts: A European Perspective on Bioethics* (Leuven: Peeters Publishers). *Euthanasia in the Low Countries* is in press (coedited with Tom Meulenbergs).

ANDREA VICINI, S.J., M.D., Ph.D., is teaching fellow of moral theology and bioethics at the Faculty of Theology of Southern Italy:

S. Luigi, Napoli, Italy. Current research projects include end-of-life issues, reproductive technologies, genetics, biotechnologies, and globalization. He is currently working on a volume on ethical issues related to genetic information. Among his recent articles are "Ethical Debate on Stem Cell Research and Roman Catholic Insights," in *Medicina nei Secoli—Arte e Scienza/Journal of History of Medicine* 15, no. 1 (2003): 71-85; and "Ethical Issues and Approaches in Stem Cell Research: From International Insights to a Proposal," in *Journal of the Society of Christian Ethics* 23, no. 1 (Spring/Summer 2003): 71-98.

1

Introduction to the Project: Genetics, Theology, and Ethics

The present collection represents the work of an international group of Catholic theologians and bioethicists who met annually for five years (1996-2001), with the support of the Porticus Foundation, to study questions of "Genetics, Theology, and Ethics." Group members were from the United States, Belgium, Germany, and Brazil and were joined by Dutch representatives of Porticus. The group met four times at the Jesuit Institute at Boston College, and once at the University of Tübingen. Also participating in these meetings were dialogue partners from the fields of sociology, philosophy, law, and genetic science. This group studying the theological and ethical aspects of genomics was one of four companion projects. The other three, on counseling and the psychosocial aspects of genetics, international law and policy, and genetic factors in Duchesne muscular dystrophy, were centered, respectively, at the Center for Human Genetics of the Catholic University of Leuven, Belgium; Center for Research on Public Policy, University of Montreal Law School; and the Center for Genetic Medicine, Children's Research Institute, Washington, D.C. All groups were interdisciplinary and brought values and concerns from the Catholic tradition into conversation with other approaches.

The Genetics, Theology, and Ethics Group was presented first with issues surrounding genetics research on diagnosis of and intervention in disease and therapy and counseling of families with affected children or those who were at risk of bearing children with genetic anomalies. Therefore, questions of immediate concern had to do with individual and family decisions about reproduction and genetic testing. The considerations of the group quickly broadened out to include social questions, such as allocation of resources and access to benefits. The group also frequently discussed problems related to the national and international regulation of research, including issues as diverse as research on embryos, genetic enhancement, and patent rights.

The papers in this volume were developed against the background of official Roman Catholic teaching, but they do not all take the same positions. They are also accompanied by brief essays by four respondents representing different fields and nationalities. As a context for attending to the resulting interdisciplinary conversation about genetics, theology, and ethics, it may be helpful to offer a concise overview of some of the ways in which religion and the ethics of genetics could be related, as well as of key aspects of Catholic teaching on genetics.

Religion, Genetics, and Ethics

Religious traditions, themes, and symbols expressing an experience of or belief in a transcendent being or power and the nature of human life in relation to the transcendent do not offer many specific guidelines for the development of genetic research or its use. They do, however, create a context of values, dispositions, and practices within which specific ethical questions can be taken up.

Perhaps the most important contribution religion can make to public debates about the applications of genetic research is to suggest a different worldview, an extraordinary orientation that engages the imagination and provokes a reconsideration of priorities. What are its elements? Key to any religious perspective is a transcendent frame of reference. Human projects and aspirations are set against a person, power, or source of existence that is higher or larger than human history, that is qualitatively different than the human, and in light of which the human must be judged. Jews, Christians, and Muslims evoke this larger reality with the biblical story of Creation. Creation, however, can be used in different ways in bioethical debates, depending on what values a thinker or community discerns as needing emphasis in a given situation. For example, some theologians join this symbol to the genetics discussion by saying that humans are made "in the image of God," and are called to be "co-creators" with God. Others, fearing human hubris and an uncritical approach to scientific innovation, use the idea that humans are finite creatures with set limits. Created nature may be seen as a God-given boundary; life as such may be portrayed as "sacred"; or scientists and their supporters may be warned against "playing God."

Such a warning reflects a religiously sensitive perception that humans are capable of misusing their created powers and of willfully doing evil. A symbol that powerfully captures this insight is one found almost universally in religion: sin. While the scientific worldview tends to be pragmatic, results-oriented, empirical, optimistic, and humanistic, the religious worldview recommends humility, prudence, caution, and even repentance. Aware of finitude and sinfulness, religious thinkers are likely to accept that some suffering is part of the human condition and to be more skeptical that human knowledge and "advancements" can relieve suffering to the degree that scientists may claim. On the other side of sin, however, religious traditions also offer hope of redemption. For the most part, this hope is eschatological—it will be fulfilled only in a world or sphere beyond history and time as we know it and only by the power of God. Still, one aim of a religious way of life is to offer some experience of redemption even in this life. In this light, science and medicine may be seen as part of the healing that humans may accomplish with divine blessing.

Selecting and interpreting religious symbols depend in significant part on the context and concerns of the interpreter and can get especially complex when goods and values conflict. For example, stem cell research may cure disease, but it also destroys a form of early human life. Cloning may relieve the pain of infertility but it introduces the new problem of creating children whose parents have much more control over their DNA than has ever been the case, children who can be thought of as having only one genetic parent—the adult whose cell provided the blueprint for their creation. Pharmacogenomics will offer much more precise and effective ways to attack disease, but it will also be incorporated into a biotech and pharmaceutical industry that is sure to channel miracle cures to those who have money or health insurance. The value of genetic therapy may very well collide with the value of equitable access, especially if the vastly unequal levels of health and disease worldwide are taken into account.

Compassion, one of the most important and universal religious values, provides a guide or at least a bias of concern in discerning the right way through conflicts such as these. In the words of Johann Baptist Metz, "Compassion is the key word for a global program for Christianity. Compassion." Compassion is described by Metz as "political empathy" that "points toward a comprehensive justice" demanding action.[1] Compassion is certainly not limited to religious traditions and believers, but it is a key religious value and virtue. Inherent in the meaning of creation and especially redemption is a call to altruism and inclusive community, where the suffering of the most vulnerable and excluded becomes a special focus of healing. Abrahamic and Christian religious symbols that communicate this priority are covenant community, prophetic care for the "widow and orphan," the commandment to love God and one's neighbor as oneself (in Deuteronomy and all four Gospels), the idea that even former enemies may be gathered into the "body of Christ," and Luke's parable of the Good Samaritan. Many religious traditions recognize a version of the "preferential option for the poor" as key to faith. Religious ideals imaginatively call our attention to the human capacity for empathy, love, and mercy and to a human calling to relieve suffering and empower the voices of the vulnerable and marginalized They also create communities of identity, motivation, and action where such ideals can be cultivated—though religious persons and groups must always repent of their own sinfulness and failure to put their ideals into practice.[2] In the words of Catholic theologian Maura Ryan, the church is "not exempt from its own pretensions or temptations to self-deception." Yet it still plays an important role in "testing the sincerity of commitments to respecting persons as subjects rather than as objects," and in "calling powerful nations such as the United States to translate the rhetoric of solidarity into economic and political commitments."[3]

In addressing the social implications of genetics, religion urges us to look at the big picture and to ask in whose interests scientific knowledge is being advanced and who is most likely to profit from it. Concerns about the ethics of cloning, stem cell research, and the development of designer drugs can all in some sense be traced to concern about those who will be harmed or left out, whether they be embryos, children, the uninsured, or populations of the developing world. As genetics and biotechnology intersect with an increasingly globalized and market-driven economy, the preferential option for the poor may seem as difficult to attain as it is urgent. Religion does not offer specific resolutions to the complex social problems that the new genetics poses. But it does demand that our moral attention, analysis, and action be informed by humility, compassion, and ideals of social justice. These virtues and ideals do not as such furnish the specific norms necessary to put commitment into practice. But they do provide the dispositions and motivations necessary to act in favor of just treatment for all. In addition, the ideal of inclusive community across class and cultural boundaries also suggests that just social relations will not result from the good will of the powerful alone, but will require the agency of those not yet part of the decision-making process about the future of genetics.

Catholic Theology and the Ethics of Genetics

Catholic tradition provides resources for looking at genetics in terms of the value of scientific progress and healing; of individual decisions focused on reproduction; and of distributive justice and the common good.

As is well known, the contemporary teaching authority of the Catholic Church (called the *magisterium*) places great value on the early embryo and forbids the direct destruction of embryos even for the most serious of reasons.[4] It is less well known that the church and John Paul II specifically have taken a positive attitude toward genetic research in general, as long as it aims to promote human welfare within the limits required by "full respect for man's dignity and freedom."[5] "A strictly therapeutic intervention, having the objective of healing various maladies—such as those stemming from chromosome deficiencies—will be considered in principle as desirable, provided that it tends to real promotion of the personal well-being of man, without harming his integrity or worsening his condition. Such intervention actually falls within the logic of the Christian moral tradition. . . ." Though he does not spell out the specific requirements, the pope does not even rule out genetic enhancement, so long as it is pursued with respect for fundamental human dignity, for the "common biological nature which lies at the base of liberty," and without marginalizing any social group.[6]

Two clear limits on genetic manipulation, according to official Catholic teaching, are the life of the embryo (in its earliest stages sometimes called the "pre-embryo"), and the inviolability of the biological procreative process (that is, the link between sex and procreation). As *Donum vitae* states it,

> The Magisterium has not expressly committed itself to an affirmation of a philosophical nature [about whether the embryo as a human individual is a human person], but it constantly reaffirms the moral condemnation of any kind of procured abortion. . . . The human being is to be respected and treated as a person from the moment of conception.[7]

Moreover, "the link between the meanings of the conjugal act and between the goods of marriage" demands that "the procreation of a human person be brought about as the fruit of the conjugal act specific to the love between spouses."[8] Consistent with these views, official statements have condemned embryonic stem cell research and cloning.[9]

In recent years, the pope has turned his attention increasingly to the social justice aspects of medicine and health care. In his social encyclicals, the disparity between rich and poor peoples and nations arises frequently as worthy of great concern. At a conference sponsored in 1998 by the Pontifical Academy for Life, John Paul II did not confine his remarks to the traditional "pro-life" agenda. While undoubtedly affirming the integrity of the procreative process and the rights of the embryo, he equally expressed the concern that "research benefits should be shared with developing nations," so as to prevent a further source of inequality. Taking note of the reality that "enormous financial resources are invested in research of this sort," which "could be allocated first and foremost for the relief of curable illness and of the chronic poverty of so many," he called on all societies to cooperate for the common good.[10]

Another example is the pope's remarks to U.S. president George Bush on stem cell research. In the seventh paragraph of an eight paragraph message, he condemns "the creation for research purposes of embryos, destined to be destroyed in the process," and defends "human life at any stage, from conception to natural death." In five of the remaining paragraphs, he specifically names social justice and the common good, alluding especially to globalization, the "fault line" between those who can and cannot benefit from the opportunities of stem cell research, the debt of poor nations, and the need for equitable access to technology. Reminding

Bush that "A global world is essentially a world of solidarity!" he calls the president to take responsibility for leadership in building "a world in which each member of the human family can flourish."[11] Although the papal reinforcement of the Catholic prohibition of destroying early human life was cited many times by the media, the exhortations suggesting that the ethics of genetics must include considerations of social justice have much more rarely been noted.

The Essays in This Volume

The present essays all appreciate the promise that genetics research holds for human health and well-being. However, distributive justice is a major concern, as is international policy aiming to guide genetic development toward the truly common good. Several authors also deal with the morality of destroying or commercializing early life and of introducing technological interventions into the procreative process itself. None of these, however, are separated from their social and political context.

The first two essays, by project members Dietmar Mieth and Paul Schotsmans (with Bart Hansen), differ on whether certain possibilities of modern genetics should be viewed positively or negatively. Hence, they illustrate the point that religious traditions and symbols may set a context for the assessment of specific alternatives, without necessarily requiring that the factors at stake and the ultimate path to be taken can or will be read in the same way and with the same results. Both explore the question of whether human use of new genetic knowledge should be seen as somehow sharing in the divine creative process. Mieth is very cautious, while Schotsmans is more trusting that new human powers can support human welfare while avoiding risks. It is interesting to know that Mieth has been involved on many national and international bioethics commissions and committees, including some established by the European Parliament and the United Nations. He is concerned about the increasingly permissive direction international law is taking, especially the fact that barriers to certain types of research are established only provisionally (in contrast to German law, which bars all research on the embryo). Schotsmans, on the other hand, is involved at his university with a bioethics center and with counselors who advise couples who have a strong desire to become parents and who face high barriers, including for some the risk of genetic disease. Options like preimplantation genetic diagnosis may help some of these couples. Therefore Schotsmans is more open to recent reinterpretations of the status of the embryo that see it as having significant value only after "individualization" at about fourteen days. This might allow more latitude for interventions regarding reproduction in such cases.

Paul Schotsmans and Bart Hansen take the images of creation, co-creation, and image of God in a proactive direction, discussing nonreproductive cloning and stem cell research. They discuss recent suggestions by theologians that embryos before individualization may have a different status than those after it. Taking up the problem of whether the attempt to exercise mastery over nature through therapeutic cloning amounts to "playing God," they grant that human createdness and free will are interdependent, and that some types of genetic manipulation could lead to a level of determinism in which freedom and creativity are almost eliminated. To settle the issue, they return to an interpretation of the symbol of creation. In their view, God is self-limiting in relation to humanity, giving creatures a "received autonomy" that allows them to be "created co-creators." This power, however, will not necessarily be utilized for the good. To reflect on the proper use of human freedom within limits, they return to positive biblical images of Jesus healing the sick.

In his chapter, then, Dietmar Mieth argues that humans can in a sense be "coagents" in the realization of God's creation, but that this can only be done ethically if the parameters set by human finitude are respected. Mieth uses literature and philosophy, especially the final act of Goethe's *Faust*, to display imaginatively the disastrous consequences of human attempts to control the outcome of events by excessive, forceful interventions. He also makes the case that "extracorporeal fertilization" undermines the human meaning of sexual relations and procreation because their separation means an unacceptable loss of physicality in the process of becoming parents. He observes that interventions are usually carried out on the bodies of women and asks whether they represent true "self-determination" if the opportunities for and limits of such choices are determined by cultural and economic factors. Choices about reproductive technologies are not just personal matters. They may result in or reinforce structural sin by contributing to the loss of the unconditional acceptance demanded by respect for persons. Mieth fears that a social ethos of "solidarity and assistance" is losing out to a social ethos in which "guilt" is assigned to individuals who do not meet standards of acceptability. This contradicts the Jewish and Christian view of history as a progressive purification of the image of God, though always in light of humanity's intrinsic finitude.

Kevin FitzGerald, a Catholic bioethicist and genetics researcher, provides a larger philosophical framework for these kinds of debate, arguing that they are always situated within "philosophical anthropologies," or "comprehensive theories of human nature." He identifies four types: static theories, governed by philosophical or theological beliefs about human nature as fundamentally unchanging; scientistic theories, which rely on science as the key to understanding the nature of the human; dichotomized theories, which separate human biology from moral characteristics or personhood, thus allowing machines, for instance, to be considered persons; and dynamic theories, which integrate scientific information with historically conditioned philosophical concepts. A model of this type of theory is the philosophical anthropology of Karl Rahner, who emphasizes human freedom, but sees it as always realized in relationships. Rahner also understands human nature, as free, to be self-constituting to a certain extent. FitzGerald illustrates the import of this theory for genetics by showing the blurred line that exists both scientifically and culturally between genetic therapy and enhancement.

James Keenan, a theological bioethicist with expertise on Thomas Aquinas, takes up the usefulness of an approach to ethics that emphasizes virtue rather than act-focused decision-making. Virtue ethics can provide an approach to genetics that is dynamic and permits reformulation of the notion of virtue itself in light of "who we can become"; that works with a social, relational self-understanding; that defines the virtuous persons relationally, in terms of justice and equality; and that is compatible with liberal societies, since it has at its center a relational, autonomous subject. In Keenan's relational anthropology, the three cardinal virtues are defined as justice, fidelity, and self-care, and correspond to identity as general or human, as specific or communal, and as unique or individual. Regarding genetics, virtue ethics offers dispositions or values of character such as respect, but these also must lead to guidelines. Since virtue ethics is oriented to the distant horizon of human becoming, it is not quick to preempt specific means with negative prohibitions. Yet it can define some means, such as genetic enhancement or germ-line therapy as imprudent, especially in view of the fact that, as FitzGerald also argues with regard to Rahner, genetics is the manipulation "of our very selves."

The final two essays, by myself, Lisa Sowle Cahill, and Márcio Fabri dos Anjos, turn to aspects of Catholic social teaching to pursue questions of justice in relation to genetic ethics. My prior writing in the areas of gender ethics and war and peace leads me to place the ethics of genetics in the context of the common good and to be wary of any undue influence of market forces in determining social relations and access to knowledge and goods. Márcio Fabri is a theologian and bioethicist writing from the standpoint of Brazil, a country in the developing world. He uses the resources of liberation theology to address the realities of power and vulnerability that determine who controls and benefits from the new genetics. My essay notes that some international policy statements, such as the 1997 UNESCO Universal Declaration on the Human Genome and Human Rights, place genetic knowledge in the context of social responsibility, using terminology such as "solidarity" and "common heritage." At the practical and policy level, however, market incentives seem more determinative of the direction of research than do global norms of social accountability. Although there is no single world oversight body to regulate genetics research, experience in other areas (women's equality, human rights, the environment) offers some evidence that transnational institutions and networks are emerging that exert pressure on and limit the influence of market incentives. Catholic social teaching about distributive justice, the common good, solidarity, subsidiarity, and the preferential option for the poor is discussed and its relevance to genetics is explored using patent law as an example.

Márcio Fabri writes more explicitly from the standpoint of the poor and the voiceless themselves. He points out that not only individuals but also groups benefit from genetics research. In Latin America, many social groups are vulnerable and lack the social conditions necessary for participation in ethical analysis of genomics. Nonetheless, Latin America is an increasingly important contributor to technological and scientific research. Brazil is the second most important country to contribute to the screening of the human genome, and has about four hundred health research centers.

Genomics depends on power over technical progress, economic investments, political capacity, and cultural symbols. Taking his cue from liberation theology and the Bible, Fabri argues that power can also be understood as empowerment, as a dynamic force for the common good, and as a sign of transformation, as well as a form of domination in contradiction to Jesus's actions. Finally, Fabri holds up an ideal notion of perfection as love and mercy, based on the narratives about Jesus's healing human imperfection, sin, and suffering. Genetics can contribute to the realization of this ideal, within a concern for human dignity. It is necessary, however, to develop local, interactive, participatory processes of analysis that involve civil society and are based on more extensive popular education.

Taken together, these essays offer many concerns about the directions genetics research is presently going, but still value the potential for relief of human suffering and improvement of human well-being in the emerging discoveries of genetic science.

Four conversation partners engage the questions raised by the theologians represented and add the perspectives of disciplines as varied as law, medicine, philosophy, and counseling psychology. Bartha Knoppers, a Canadian and an expert on international law and social policy, observes that the essays in the book reflect the complexity of the subject matter. She proposes that a dynamic ethical dialogue would make it possible for genetic science to advance both human freedom and responsibility as well as to serve as an instrument of social justice. Andrea Vicini is an Italian moral theologian as well as a medical doctor with a specialization in pediatrics. While applauding the ethical concern with universality that pervades the present essays, he also raises the question whether the issue of particularity may need to become more central in the future. Genetic science will not only be appropriated in different contexts, it may also reveal some of the particular genetic traits that vary from population to population, hence requiring different approaches to disease. Gerry Evers-Kiebooms heads a center in Leuven, Belgium, for psychosocial counseling of those at risk for genetic disease. She brings some very specific and practical concerns about how to handle and communicate genetic information and the ambiguity of most predictive genetic testing. Finally, Hasna Begum is a women's rights and health care rights advocate in Bangladesh, as well as a philosopher. From her experience in the developing world, as well as in international bioethics debates, she appreciates many points of the current volume but also presses further questions about global justice in health care and biomedicine.

Notes

1. Johann Baptist Metz, "Toward a Christianity of Political Compassion," in *The Love That Produces Hope: The Thought of Ignacio Ellacuria*, ed. Kevin Burke, S.J. and Robert Lassalle-Klein (Collegeville, Minn.: Liturgical Press, forthcoming) manuscript p. 2. See also Hille Haker, "Compassion as a Global Programme for Christianity," *Concilium* 2001/4 (2001): 55–70. Haker discusses Metz, and locates the foundation of compassion in the "negative universalism" of human suffering.

2. For views of the role of religion in bioethics that are similar to the one expressed here, see Dena S. Davis and Laurie Zoloth, eds., *Notes from a Narrow Ridge: Religion and Bioethics* (Hagerstown, Md.: University Publishing Group, 1999); and Elliot N. Dorff, "A Narrow Ridge, A Larger Vision," *Hastings Center Report* 31 (2001): 44–46.

3. Maura A. Ryan, Introduction to *The Challenge of Global Stewardship: Roman Catholic Responses,* ed. Maura A. Ryan and Todd David Whitmore (Notre Dame, Ind.: University of Notre Dame Press, 1997), 7.

4. An important recent statement of this position is the 1987 document of the Congregation for the Doctrine of the Faith, *Donum vitae* (*Instruction on Respect for Human Life in Its Origin and on the Dignity of Procreation: Replies to Certain Questions of the Day*) *Origins* 16 (1987): 697, 699–711. The primary purpose of this document was to condemn reproductive technologies like in vitro fertilization.

5. John Paul II, "Biological Research and Human Dignity," address to the Pontifical Academy of Sciences, *Origins* 12 (1982): 342–43.

6. John Paul II, "The Ethics of Genetic Manipulation," address to the World Medical Association, *Origins* 13 (1983): no. 6.

7. *Donum vitae*, no. 1.

8. Ibid., no. 4.

9. Pontifical Academy for Life, "Declaration on the Production and the Scientific and Therapeutic Use of Human Embryonic Stem Cells," *L'Osservatore Romano,* weekly edition in English, September 13, 2000, 7; available from the website of the Eternal Word Television Network, www.ewtn.com /library/CURIA/PALSTEM.

10. "Discourse of Holy Father John Paul II," in *Human Genome, Human Person and the Society of the Future: Proceedings of the Fourth Assembly of the Pontifical Academy for Life,* ed. J. Correa and E. Sgreccia (Vatican City: Libreria Editrice Vaticana, 1999): 8–9.

11. John Paul II, "Remarks to President Bush on Stem Cell Research," *The National Catholic Bioethics Quarterly* 1 (2001): 618.

PART I

PROJECT CONTRIBUTIONS

Section I: Fundamental Orientations

2

Stem Cell Research: A Theological Interpretation

BART HANSEN AND PAUL SCHOTSMANS

Certain events settle themselves in the collective memory of humankind where they keep functioning for decades as points of reference for future generations. The announcement of the successful cloning of Dolly is such an event. Every one of us will remember this thought-provoking occasion or will, at least, have been confronted with the extended media coverage of this breakthrough in medical science. Immediately, world leaders reacted, and the question was raised how long it would take before the shepherd himself would be cloned.[1] At the same time, Dolly has narrowed the gaze of the world's ethicists and theologians, for by a fixation on her, they have become hyperfocused on one technique—reproductive cloning. More important is the sequel to this medical breakthrough: the successful Dolly experiment led to the possibility of human embryonic and adult stem cell research.[2]

We first want to clarify the technological possibilities in an age of biological control. A careful terminology will help public discourse. Second, we will offer a short overview of the most recent opinions from national ethics committees or similar bodies, public debate, and national legislation in relation to human stem cell research. Third, we will evaluate stem cell research by means of two ethical methods: the clinical-ethical approach and the cultural-philosophical one. The former will concentrate on problems discussed in medical ethics committees, whereas the latter will focus on a specific cultural theme, that is, "playing God." Finally, we will reflect on the Christian theological tradition and will try to indicate how ethical discourse on biotechnological issues is informed and shaped by specifically theological themes. To some, it seems that humankind has usurped the Divine Creator's place and, by doing this, has crossed a border. To others, this medical breakthrough is in line with previous technological innovations in medicine. Our stand is essentially close to this more open and optimistic perspective: the human being as a *created co-creator* is in our opinion not standing up against God, but he is taking full responsibility in realizing God's intentions.

A Terminological Clarification

Some terminological clarifications could help us better understand the issues at stake. Cloning refers to a clone, from the Greek word for twig, and in extension, to cutting and grafting. In science it originally referred to a neutered animal or plant. In the actual scientific jargon, the word "clone" is used to specify a genetically identical copy of a gene, a molecule, a cell, a plant, or an animal. "Cloning" refers to, if applied to an organism, the production of one life form or a group of living beings that share a number of genes identical to the genes of the organism that lay at the basis of the reproduction. The production of completely identical organisms is an everyday practice in plant breeding where cloned organisms are commonly referred to as "varieties." The reproduction of numerous important plant varieties is carried out starting from small cuttings of other plants. In zoology, this form of reproduction is only used with minimally evolved species. In the case of vertebrates, the occurrence of identical twins is a spontaneous form of cloning. Monozygotic twins come into being by an embryo splitting during one of the earliest phases of development. Because they originate from one zygote that resulted from the fertilization of one ovum by one sperm cell, monozygotic twins are identical. Nevertheless, they differ from their parents.

Molecular cloning refers to a routine technique in molecular biology that consists of cloning the molecular basis of heredity, the DNA. DNA fragments are copied and amplified in a host organism, usually a bacterium. This technique has led to the production of such important medicines as insulin, growth hormones, erythropoietin (necessary to treat anemia associated with dialysis for kidney disease), and tissue plasmogen activator (tPa) to dissolve clots after a heart attack. In cellular cloning copies are made of cells derived from the soma, or body, by growing these cells in culture in a laboratory. The genetic makeup of the resulting cloned cells, called a cell line, is identical to that of the original cell. This, too, is a highly reliable procedure, which is also used to test and sometimes to produce new medicines, such as those listed above.

In order to comprehend the differences between the scientific procedures relied on to clone organisms, blastomere separation will be distinguished from somatic cell nuclear transfer. Furthermore, the reality behind the concept of so-called therapeutic cloning (as opposed to reproductive cloning) will be unveiled, with reference made to stem cells.

Cloning by Blastomere Separation

Cloning by means of blastomere separation is an artificial process of what occurs naturally as twinning. Cells of an embryo created by means of "old-fashioned" sexual reproduction—the fusion of an ovum with a sperm cell—are separated from one another in the two- to eight-cell stadium. Each cell, called a blastomere in this early phase, is able to produce a new individual since blastomeres are totipotent. The embryos and organisms thus produced are identical to each other, though not identical to the donors of the gametes, that is, the parents. At the October 1993 Congress of the American Fertility Association, an attempt to produce human twins by means of blastomere separation was presented.

Cloning by Somatic Cell Nuclear Transfer

Cloning by means of somatic cell nuclear transfer (SCNT) consists of replacing the ovum's haploid nucleus by a diploid one coming from a differentiated somatic cell. With this type of cloning, there is only one genetic parent—the donor of the nucleus. The experimental procedure, which led to the birth of Dolly, can shortly be described as follows: First, cell cultures were derived from the mammary gland of an adult sheep. They starved the donor cell line by removing all nutrients from the medium prior to nuclear transfer. Under these starvation conditions, the cells exit the cell cycle and enter the so-called G0 state (Gap phase 0), in which chromosomes have not replicated. Fusion of G0 nuclei with eggs ensures that the donor chromosomes have not initiated replication prior to fusion. From a second female sheep, eggs were isolated and enucleated. Then, electric impulses induced a merger of mammary gland cell and an enucleated egg, thereby reactivating the genes of the mammary gland so that they could control the growth of the fused cell. Later, the growing embryo was implanted in the uterus of a third sheep, and in the fifth month, all this gave rise to Dolly, a lamb identical to her genetic mother. To attain this successful birth, 277 attempts were undertaken.[3]

Reproductive cloning is still to this day an inefficient and error-prone process that results in the failure of most clones during development. Most clones die soon after implantation. Those that live to birth often have common abnormalities irrespective of the type of donor cell used. For instance, newborn clones are frequently unusually large and have an enlarged placenta (the large-offspring syndrome). A large proportion of neonatal clones also suffer from respiratory distress and/or defects of the kidneys, liver, heart, and brain. Even long-term survivors can have abnormalities later in life. Aging cloned mice were recently reported to become obese, die prematurely, and have tumors.[4]

TABLE 1
COMPARISON OF CLONING TECHNIQUES

	Somatic Cell Nuclear Transfer	Blastomere Separation
Number of Genetic Parents	1	2
Degree of Resemblance with Genetic Parents	High	Low
Number of Animals	Indefinite (theoretically)	Limited

Therapeutic versus Reproductive Cloning

Therapeutic cloning is a compound concept that can cause confusion. Given the similarity between therapeutic and reproductive cloning techniques (both use SCNT), there is a methodological problem concerning the term "cloning." To distinguish between both, one has to gain insight into the researcher's intentions. The difference between reproductive and therapeutic cloning implies that, for the latter, one limits the process to the development of the embryo in vitro and, thus, does not proceed to the implantation in the uterus for reasons of giving birth to a child. In light of these similarities in means, we prefer to speak about "the" cloning technique with either a reproductive or a therapeutic end.[5]

Cloning for therapeutic purposes opens up new and interesting perspectives for transplantation medicine, where tissue compatibility is of vital importance. For the interest of the scientific community the difference between stem cell research and human reproductive cloning must also be clearly articulated. Scientists themselves try to find a more appropriate term for the use of somatic cell nuclear transfer to create stem cells. They propose the term "nuclear transplantation" (table 2).[6] According to them, this term captures the concept of the cell nucleus and its genetic material being moved from one cell to another, as well as the nuance of "transplantation," an objective of regenerative medicine. The embryo that results from the combination of nucleus and enucleated ovum that launches SCNT is called by some "clonote," instead of zygote, that is, the first stage of an embryo produced by fertilization.[7]

TABLE 2
THE CRUCIAL DIFFERENCES

	Nuclear Transplantation	**Human Reproductive Cloning**
End Product	Cells growing in a petri dish	Human being
Purpose	To treat a specific disease of tissue degeneration	Replace or duplicate a human
Time Frame	A few weeks (growth in culture)	9 months
Surrogate Mother Needed	No	Yes
Sentient Human Created	No	Yes
Ethical Implications	Similar to all embryonic cell research	Highly complex issues
Medical Implications	Similar to any cell-based therapy	Safety and long-term efficacy concerns

In order to grasp fully the therapeutic purposes adequately, we also need to clarify the most important means to these ends, that is, stem cells. Stem cells are cells that can divide to produce either cells like themselves (immortality), or cells of one or several specific differentiated types (potentiality). Stem cells can be found throughout every stage of human development—from embryo to adult—but their potentiality decreases with age: embryonic stem cells (ES cells) are considered to be pluripotent, whereas adult stem cells are less versatile; they are multipotent.

The advantage of using ES cells in cloning procedures is twofold. On the one hand, ES cells are most promising as they retain the greatest *potential* to develop into a wide range of tissues. On the other hand, their so-called *immortality*—their capacity to undergo prolonged undifferentiated proliferation—gives researchers ample time to add or delete DNA precisely. Thus theoretically, a human ES cell could be genetically altered. Then, by means of somatic cell nuclear transfer, the genetically altered ES cell could be fused with an enucleated egg to create an embryo. The embryo's inner cell mass is made up of ES cells that would then give rise to transplantable tissue: therapeutic cloning has occurred.

In the next paragraphs the advantages and disadvantages of both human embryonic and adult stem cell research will be analyzed. However, before examining this matter thoroughly, we want to indicate that a different type of cell nuclear replacement could be used to help women avoid the birth of a child with inherited mitochondrial diseases. Mitochondria are small energy-producing structures in the cytoplasm of every cell. The cytoplasm can be thought of as a jelly, which holds the nucleus of the cell. Although the vast majority of the DNA is contained in the nucleus of the cell, the mitochondria also contain DNA. We now know that mitochondrial DNA affects a number of important functions related to the role of mitochondria in providing energy for the cell. Tissue with high demands for energy, such as muscle, heart, brain, and eye are particularly vulnerable to mitochondrial defects. There are more than fifty inherited diseases of metabolism that are known to be caused by defects in mitochondrial DNA. A baby inherits mitochondrial DNA only from its mother because mitochondrial DNA in the sperm does not appear to pass through the process of fertilization. If the maternal mitochondrial DNA carries a disorder then it will always be passed on to the child. It may become possible to prevent the child from inheriting the disease by using the cell nuclear replacement technique. This would involve inserting the nucleus of the mother's egg into a donor egg that has healthy mitochondrial DNA and which has had its nucleus removed. This new egg could then be fertilized by the sperm of the woman's partner. Any child born would have received its nuclear DNA from its mother and her partner but would have healthy mitochondrial DNA from the donor egg.[8]

Human Embryonic Stem Cell Research

Stem cells are classified into two groups, embryonic and adult (also called "somatic") stem cells. *Embryonic stem cells* can be obtained directly from embryos, from the fetal germ line, or by procedures such as somatic cell nuclear transfer. *Adult stem cells* can be obtained from fetal organs and tissues, placental blood or organs, and tissues from adults and children.

The Belgian Consultative Committee on Bioethics states in an internationally accepted definition: "Embryonic Stem Cells (ES-cells) are cells deduced from the inner cell mass of the blastocyst, which will form the fetus. Those cells are pluripotent and can differentiate into all cell types of a human adult."[9] We will try to analyze this definition in the subsequent paragraphs.

A possible approach to human stem cell research would be to derive stem cells from very early embryos (preimplantation embryos). About five to six days after fertilization, embryonic stem cells can be taken from the inner cell mass of the blastocyst. The blastocyst begins to form as a fifteen- to twenty-cell cluster just beginning to separate into identifiable parts that will go on to form the placenta, fetus, and other associated tissues. A blastocyst can be compared to a hollow ball of cells. The outer layer of the blastocyst's cells goes on to form the placenta, while the inner cells form the embryo and its membranes. Blastocysts used for the isolation of stem cells would probably have between 150 and 200 cells. After the blastocyst stage the opportunity to extract stem cells is gradually lost as the stem cells start to become specialized and no longer have the potential to become any type of tissue.[10]

Research suggests that human embryonic stem cells can give rise to many different types of cells. They raise the possibility, therefore, of major advances in health care. For example, ES cells could be used to generate replacement cells and tissues to treat many diseases and conditions. This treatment would be especially important for those cells that can no longer be renewed by an adult body. For example, we could produce heart muscle cells to treat heart disease, nerve cells for the treatment of Parkinson's or Alzheimer's disease, insulin producing cells to cure diabetes, bone marrow for leukemia patients, and so on. In short, if ES cells could be directed to differentiate into particular tissues and be immunologically altered to prevent rejection after engraftment, they could lead to the treatment or cure of the major lethal diseases of this century: neurodegenerative disorders, cancer, and heart and coronary disease.[11]

Although ES cells are scientifically promising, their isolation and growth in culture also raise ethical problems. Human embryonic stem cell research is inextricably bound to the manipulation of human embryos. The major point of controversy is that embryos are being destroyed in the course of ES cell research. The *Convention on Human Rights and Biomedicine* of the Council of Europe—the most important European consensus document, though not yet ratified by several European countries—explicitly prohibits the creation of embryos for research purposes (article 18, part 2).[12] This is one of the reasons why several research projects plan at the moment to use the spare cryopreserved embryos after In Vitro Fertilization (IVF).

Adult Stem Cells

The current technical obstacles of human cloning technology combined with the ethical controversy about using embryos for research purposes have both stimulated the scientific discovery of "embryo-saving" alternatives, that is, adult stem cells.[13]

Scientists have known for decades that certain kinds of stem cells lurk in adult tissues, for example in skin. Wounded skin heals out of a special cell layer in the epidermis, which has the capacity to cover the wound and to heal fast. One could consider cultivating cutaneous stem cells to treat skin covered in burns. Until recently most scientists have assumed that the adult-derived cells have a limited repertoire. Just as years of training usually commit a concert violinist to a career in music, so scientists have assumed that when a young cell takes on an identity and turns various suites of genes on or off that genetic programming irreversibly commits it to becoming one of just a few cell types. In 1999 the *British Medical Journal* reported that signals in the immediate environment could sometimes overwrite a cell's genetic history, implying that adult stem cells will receive the versatility of an embryonic stem cell (that is, pluripotency).[14] For instance, stem cells isolated from the bone marrow of an adult patient could become blood, brain, muscle, cartilage, and bone cells.[15]

Using adult stem cells for transplantation medicine increases human tissue compatibility and immune tolerance, because donor and patient are one and the same person. But even adult-to-adult cell therapy does not liberate us from the ethical and legal entanglements surrounding stem cell research, that is, the moral status of the human embryo. Reprogramming an adult stem cell to the pluripotent state of an embryonic stem cell requires further research on the molecular regulation and differentiation that can only be reached by cultivating embryonic tissue. In other words, the simultaneous development of both research strategies—adult and embryonic stem cell research—is preferable, considering that research on ES cells probably will contribute to speeding up and optimizing clinical applications of adult stem cells. These reflections may lead one to question the validity of opinions such as the opinion of the Pontifical Academy for Life, which condemns ES cell research while promoting research on adult stem cells.[16] Such an approach relies on the scientific potential of alternative options. It is justifiable only if adult stem cells prove to have the same broad applicability as ES cells from preimplantation embryos, which even now is still unlikely.

Although we have to mention the promising research results obtained by the Belgian researcher Catherine Verfaillie at the University of Minnesota. She found a stem cell in the bone marrow of adults that can turn into every single tissue in the body, the so-called Multipotent Adult Progenitor Cell (MAPC). Until then (January 23, 2002), only stem cells from early embryos were thought to have such properties. If the finding is confirmed, it will mean cells from our own body could one day be turned into all sorts of perfectly matched replacement tissues and even organs. If so, there would be no need to resort to ES cell research and therapeutic cloning.[17]

This brief scientific exposé enables us to gain a better understanding of the science behind the concepts used and to examine the societal reactions that followed these scientific breakthroughs with a clearer comprehension of the issues at stake.

TABLE 3
COMPARISON OF STEM CELLS

Specialization	*Pluripotent*	*Multipotent*
Source		
Embryonic	• Embryonic stem cell • Embryonic germ cell • Somatic cell nuclear transfer	
Adult	• Multipotent Adult Progenitor Cell	• Fetal organs and tissues • Cord blood • Organs and tissues of children and adults

Societal Reactions

The rapidity with which international and regional organizations have condemned the possibility of reproductive human cloning led to the conclusion that a restrictive approach was promoted; later decisions and debates regarding therapeutic cloning have nuanced that affirmation. Shortly following the press release announcing the Dolly experiment, the United Nations Educational, Scientific and Cultural Organization's (UNESCO) General Conference adopted the *Universal Declaration on the Human Genome and Human Rights.*[18] After recalling general principles on the universal protection of human rights and democratic ideals, the text starts with the acknowledgment that human genome research and its applications open immense perspectives for the improvement of humankind in general and for human health in particular. However, the text's corpus is mainly devoted to the limitations and the risks of this kind of scientific research. Under the terms of its article 11, the declaration specifies that "practices opposed to human dignity, such as the reproductive cloning of human beings, are not tolerable."[19] Therefore, states are invited to take the necessary measures to prohibit such experiments.

On January 15, 1998, the European Parliament (EP) adopted a resolution that urges member states to draft legislation criminalizing human cloning. At the same time, the EP invited the member states and the European Union to work toward a worldwide and clearly sanctionable ban on the cloning of human beings.[20] This resolution reaffirms the position of the European Parliament as expressed in its 1993 and 1997 resolutions. Furthermore, it should be added that in 2000, in reaction to the U.K. government decision to allow research on embryos created by cell nuclear transfer, the European Parliament adopted yet another resolution on human cloning. In that latter text they condemn the position of the U.K. government while insisting on the urgency for member states to implement appropriate legislation prohibiting any form of cloning. The European Parliament's repeated calls for a ban on any type of human cloning—if not at the international level, at least at the national level—are based on the absence of conceptual difference between cloning for therapeutic or reproductive purposes and the promises of alternative methods of research (adult stem cells, fetal stem cells). The parliament considers cloning, in any form, to be contrary to morality and *ordre public*. The parliament is also reluctant to allow stem cell research or embryonic research on spare embryos, inviting researchers to limit the number of in vitro fertilized embryos.[21]

Unlike the above mentioned texts, which, though adopted by respected institutions, have no binding force, the *European Convention on Human Rights and Biomedicine* and its additional protocols are directly applicable in the ratifying countries, and thus bear close examination.

On January 12, 1998, an Additional Protocol to the *Convention of Human Rights and Biomedicine* was signed in Paris.[22] This protocol, drawing on article 18(2) of the convention, which forbade the creation of embryos for research purposes, banned any form of human cloning (article 1[1]). A group of countries politically and/or sociologically close to Belgium, including France, Luxemburg, the Netherlands, Denmark, Sweden, Finland, Portugal, Spain, Italy, and Greece, has signed this protocol.

In a thorough review on the issue of human reproductive

cloning, the Belgian Consultative Committee on Bioethics analyses the scientific, legal, and ethical arguments used in the cloning debate. In its concluding remarks, addressed to the Belgian Government in the form of an opinion, its members unanimously recommended a clear ban on human reproductive cloning. They further promote publication and debate on the psychological, philosophical, medical, and ethical aspects of the issue in order for the public to remain fully informed. They finally note that should a human clone come to life, its dignity, human nature, or legal personality should not be challenged.[23]

Whereas, as evidenced by the several documents analyzed, a consensus on the nondesirability of allowing cloning for reproductive purposes has arisen, the attitude toward cloning for therapeutic ends appears more moderate—though still controversial. The acceptability of such procedures rests on the promises of current scientific research on embryos and embryonic stem cells: the expectation that such embryos and ES cells could be used to improve IVF therapy as well as be invaluable in transplantation and regenerative medicine (see above, Cloning by Somatic Cell Nuclear Transfer, pp. 17-18).

At the international level, under the impetus of both France and Germany, the United Nations General Assembly resolved to work toward the drafting of a binding treaty with regard to cloning.[24] Although an ad hoc committee was created and met in February 2002, and a draft international convention on the prohibition of all forms of cloning was proposed,[25] the UN decided to delay further discussions on the issue in light of the controversy and lack of consensus regarding the scope of the convention. Whereas some countries, France and Germany included, were in favor of reproductive and research cloning to be considered in two separate documents, other nations promoted an all-encompassing document.[26]

Regionally, the European Parliament, despite its earlier prohibitory declarations on all forms of cloning, has explicitly rejected a move to ban human therapeutic cloning in the European Union. Only 37 of the 391 Euro-MPs voted in favor of the "Fiori report" on the social, legal, ethical, and economic implications of human genetics (November 29, 2001). Similar voices could be heard at the conference of the Life Sciences High Level Group of the European Commission, entitled "Stem Cells: Therapies for the Future" (December 19, 2001).[27] The opinion of the European Group on

Ethics in Science and New Technologies, addressed to the European Commission, although allowing stem cell research on supernumerary embryos, prohibits therapeutic cloning.[28]

Nationally, therapeutic cloning has been authorized with limits in a few countries, whether explicitly or implicitly. Other countries, prohibiting both reproductive and therapeutic cloning, have given preference to the use of spare embryos.

In February 2002, a select committee of the house of lords decided to grant scientists in the United Kingdom permission to create and carry out research on human embryo clones. Licenses for the use of cloned embryos are, however, awarded on an exceptional basis, that is, the presence of a clear and demonstrable justification that the research contemplated cannot be carried out by using spare embryos after an IVF treatment.[29] Furthermore, promoting stem cell research and the United Kingdom's scientific leadership, the U.K. government has provided its support to the establishment of the first stem cell bank, which opened on May 29, 2004.[30]

The recent Belgian Embryo Law accepts all types of research directed at therapeutic purposes and at increasing medical knowledge. This includes (implicitly) therapeutic cloning and the development of ES cell lines.[31]

Several other countries of mainland Europe have also taken a position on stem cell research and therapeutic cloning. France[32] and the Netherlands[33] have recently adopted a more moderate position on ES cell research. They oppose the creation of embryos by means of nuclear transplantation solely for research purposes. Embryos derived from IVF that are no longer required for a parental project, on the other hand, can be used for research. As for Germany, the National Ethics Committee circumvented the rigorous "embryo-law," deciding to recommend the supervised import of human stem cells.[34]

In the United States too, therapeutic cloning remains a controversial issue. However the current debate mainly focuses on the question of federal funding. On August 9, 2001, the president (George W. Bush) limited funding to research on stem cell lines derived prior to his announcement.[35] For the time being, only research on sixty-four stem cell lines originating from supernumerary embryos is federally funded.

This selective review of current international, regional, and national positions[36] regarding ES cell research and cloning—notably the absence of a consensus on therapeutic cloning—is integral to the underlying ethical debate.

The Ethical Debate

Terminological Ambiguity Leads to Ethical Obscurity

It is feared that, once the human cloning technique is optimized through research on human embryonic stem cells, scientists will only be one small step away from growing full human beings (reproductive cloning) instead of human tissues (therapeutic cloning). We have already mentioned that we prefer to speak of *the* cloning technique, whether used for reproductive or therapeutic purposes.[37] Indeed, the ambiguity in the taxonomy used could obscure the ethical debate as the distinction lies in the intention and not the technical scientific procedure. The notion of "therapy" implies a goal: to treat or to cure, as in the Latin proverb *medicus curat, natura sanat.* The ethical problem lies in the fact that such a therapeutic goal—the origin and eventual treatment of a (genetic) disease—can only be attained through controversial research on human embryonic stem cells, which, in turn, implies the subsequent destruction of human embryos. Advocates of the inviolable integrity of each individual human life, including an embryo, cannot however agree with the "healing" purpose of human stem cell research.

The ethical debate surrounding human embryonic stem cell research could therefore be compared to a Faustian bargain: the possibility of curing lethal diseases versus the instrumentalization of the human embryo. Can one turn a blind eye on embryonic research with embryos for the sake of the development of stem cell research? To answer this question we need to examine the moral status of the embryo as well as the ethical justification for banning cloning when undertaken for reproductive purposes.

The Moral Status of the Human Embryo

The nature and the moral status of the human embryo are vast and difficult subjects in ethics in general and certainly, as the focus of our paper, in Roman Catholic moral theology. The idea of the inviolable integrity of a human embryo has led the Congregation for the Doctrine of the Faith to repeatedly condemn the adoption of an instrumental attitude toward the origin of human life (*Donum vitae*, 1987; *Evangelium vitae,* 1995). While the position of the official magisterium is univocally condemnatory, it must be acknowledged that the debate on the moral status of the human embryo inside the Roman Catholic community remains open. As early as the beginning of the seventies, Karl Rahner mentioned "the uncertain rights of a human being whose very existence is in doubt."[38] Recently, a growing number of Catholic moral theologians including R. A. McCormick (U.S.A.), P. Verspieren (France), and J. Mahoney (U.K.) do not consider the human embryo in its earliest stages to constitute an individualized human entity enjoying the inherent potential to become a human being. They give the impression that they accept the distinction between genetic and developmental individualization. The former is certainly present from the earliest beginnings of life, the latter is not: "Developmental individualization is completed only when implantation has been completed, a period of time whose outside time limits are around fourteen days."[39] Those who make this case seem to prefer a return to the centuries-old Catholic position that a certain amount of development is necessary in order for a *conceptus* to warrant personal status. Embryological studies now show that fertilization is in itself a process, not a moment, and provide warrant for the opinion that in its earliest stages (including the blastocyst stage, when the inner cell mass is isolated to derive stem cells for the purpose of research) the embryo is not sufficiently individualized to bear the moral weight of personhood.[40]

This more liberal religious view on the status of early human life implies a more open attitude regarding research on supernumerary human embryos before the moment of individualization, the so-called preimplantation embryos. This viewpoint has also brought more diversity into the Catholic approach to in vitro fertilization, preimplantation genetic diagnosis, human gene therapy, and obviously also to human embryonic stem cell research. Thus, the above mentioned theologians, while opposing the creation of human embryos *solely* for research purposes, do not condemn the use of "spare embryos" from an infertility treatment. These embryos, which are normally discarded, may instead be donated for research.[41]

Although one may question whether there is a fundamental moral distinction between using supernumerary embryos in research and the creation of embryos for research purposes, one should also be reminded that techniques like IVF would not have been possible without research on embryos created solely to enhance the success rate of such infertility treatments.[42]

The issue of the creation and utilization of human embryos for research purposes (i.e., stem cell research) has not only been the focus of internal theological discussions but has also become a concern for the scientific community. To further the development of scientific knowledge while circumventing the ethical controversy of the moral status of the human embryo, other scientific paths are being explored. Two research projects could, if their results are confirmed, bring further arguments to the opponents of ES cell research. First, as already mentioned, promising research results were obtained by the discovery of the Multipotent Adult Progenitor Cells (MAPCs; see above, p. 23). Second, another ethically attractive option was reported on February 1, 2002 by scientists of Advanced Cell Technology Inc. They claim to have isolated the first stem cell lines from monkey *parthenotes* (parthenogenesis is a Greek word for virgin origin), embryos grown from unfertilized eggs. As with other primates, human *parthenotes* would not be capable of developing to full-term babies. If researchers could find a reliable way to derive stem cells from human *parthenotes*, they could avoid therapeutic cloning, that is, the creation of potentially viable embryos as a source of stem cells and their subsequent destruction.[43] However, this latter technique could probably trigger a new ethical debate, not about the status of the embryo, but rather about what we mean by a human life-form.

Reproductive Cloning: Crossing a Boundary?

Although the focus of our discussion is on embryonic stem cell research, it is useful to review briefly the main concerns surrounding cloning for reproductive purposes to understand the consensus regarding its unethical nature. First, the procedures involved entail many medical risks. As cells divide and organisms age, mutations in the DNA inevitably occur and accumulate with age. Sporadic somatic mutations in a variety of genes can predispose a cell to become cancerous. The risks of such events occurring following nuclear transfer are difficult to estimate. Second, cloned children could be exposed to unnecessary psychological suffering and social pressure or alienation detrimental to their development, specifically their sense of identity. Indeed, such children would be subject to overwhelming societal and scientific expectations and constant attention. Finally, with reproductive cloning, the shadow of eugenics reappears: can we decide and, if so, on what grounds do we decide which human traits and characteristics should be favored? An analogy can be drawn with the issue of sex selection. Similar to sex selection, cloning practiced for reproductive purposes, aiming to interfere with fundamental human traits, would undermine human freedom, notably the freedom of genetic speech. The Kantian categorical imperative that we treat "humanity in your own person or in the person of any other never simply as a means but always at the same time as an end" is applicable in this context.

Moreover, even if we could uncover optimal selection methods, reproductive cloning guarantees neither a particular phenotypical manifestation of the genes nor a carbon copy of behavioral traits, for example, character, personality, intelligence, and the like. After all, little is known about the complex correlation between genes and their environment.

The scientific "cloning" discoveries of the last few years have not only generated a heated ethical debate mainly centering on the status of the embryo but also nourished the cultural-philosophical debate.

The Cultural-Philosophical Debate

In the cultural-philosophical debate surrounding genetics and more particularly stem cell research, the discussion has mainly been articulated in terms of the acceptability for human beings to make use of the knowledge and capacity provided by scientific advances to bend God's will; of the shift between the alterable and inalterable traits and the attempts, absent theological references, to moralize human nature; and finally of gene utopia or myths.

Playing God

Many of the issues raised by stem cell research have significant theological components and therefore require a religious and theological response. For people of faith, the impact and implications of these new medical developments cannot be addressed without referring to fundamental theological precepts about human nature, standing before God, and the role of humanity in creation. Rabbi Elliot Dorff lists three theological commitments that are shared by the Jewish and Christian tradition:

- Our bodies are not ours; they belong to God and God commands that we seek to preserve life and health
- All human beings, regardless of their abilities or disabilities, are created in the image of God and are therefore to be valued as such
- Humans are not God; we are finite and fallible, and this fact ought to promote humility and urge caution.[44]

In brief, the phrase "playing God" has come to be used to convey the idea that it is inappropriate for humans to change the way other living organisms (including human beings) are constituted because it amounts to usurping the creative prerogative of God.[45] Theology of Creation seems, however, to be seriously challenged to answer these new scientific developments, if such an answer can be given. Analyzing the meaning of the phrase "playing God," the theologian Ted Peters argues that its primary role is to serve as a warning and that it has very little cognitive value when looked at from the perspective of a theologian.[43]

The expression "playing God" can, according to Ted Peters, have three distinct meanings that, in our point of view, merge in the reflection on human ES cell research. According to the first meaning, the expression pertains to the steadily growing base of human knowledge of the foundations for God's creation of the human being, namely with regard to embryonic stem cells. The laboratory has become the place where we gain insight into God's awesome secret. The Torah is no longer the lamp to light the (human) path (Psalms 119, 105); rather, the microscope has become the beacon. A small peephole, henceforth, allows us a nearly panoramic view of the so-called germ at the beginning of human life or the stem cell.

This cell, whose nucleus is protected by a cell wall, is now the new sacred place; whoever enters it treads the path of the Holy of Holies. This sacrosanct place no longer contains two stone tablets, as did the Ark, but the two new laws of life, the DNA strings. The second acceptation of the phrase points to the increased power of medicine over life and death. This power, acquired via scientific advances, has recently culminated in human ES cell research. Stem cells—to which some accord the status of the inception of human life—are considered to be the ultimate means of saving the life of patients with serious ailments. Finally, "playing God" is explained in a more or less literal manner: human beings attempt to create—as God—new human tissue or even life with the aid of (therapeutic) cloning technology. In brief, the mastering of (human) nature through (therapeutic) cloning raises the question of whether the human being, as the image of God, is permitted to carry out this task or whether God alone may exercise this right?

To answer these questions, we ought to fall back on the umbrella structure of human moral experience. That structure is based on the distinction between (1) that which we are responsible for—individually or collectively (*nurture*); and (2) that which has been given to us as a background against which we act and which cannot be altered by ourselves (*nature*). Ancient philosophy already distinguished the human capacities—handled by men themselves—from their destinies, which are in divine hands. Christian moral theology also distinguishes between the world created by God—our natural condition as creature included—and the scope of human freedom. Medical researchers use scientific language to come to the same distinction: cells and genes given by Mother Nature versus what can be done to manipulate human nature. The relationship between "createdness"—whether by God or by a natural process—and free will constitutes the backbone of moral reasoning and any alteration in this relationship is worrying.[47]

It is interesting and unsettling to study information published in the media where the use of terms like human "breeding," "selection," and "manipulation" always incite deeply rooted moral reservations. An example of a public discourse that shocked different individual as well as collective self-understandings, worldviews, and basic moral convictions was the "Sloterdijk-debate."[48] This debate caused an earthquake in the German media in the summer of 1999, but the tremors expanded to other European countries (France and the Netherlands) through 2001. The starting point of the debate was the misinterpretation by journalists of a concept used by Peter Sloterdijk, a German philosopher, in the course of a presentation entitled "Rules for the Human Breeding Ground." This presentation was part of a symposium on the theme "Exodus from Being," which was held in July 1999 in Schloss Elmau, Germany.[49] The epicenter was Sloterdijk's suggestion that, with the development of biotechnologies, human breeding was henceforth achievable by means of so-called anthropo-techniques. By anthropo-techniques he referred to a transformation of the human being by means of biotechnology, for example, preimplantation genetic diagnosis, human cloning, or ES cell research. Instead of *receiving* an identity (createdness), humans are thus able to *make* their identity by genetically modifying their bodies (self-creation). Some interpreted the eventuality that humans can "make" their identity by means of biotechnology as a substitute for the complex socialization process, an idea that explains the highly emotional character of the debate.[50]

According to Jürgen Habermas, the issue at stake is a de-differentiation not *of* stem cells to their virgin condition but *through* stem cell technology of deeply rooted Aristotelian categorical distinctions, that is, between the subjective and the objective, between the naturally grown and the humanely made.[51]

Therapeutic cloning or human embryonic stem cell research seems to disturb this moral balance. The prospect of designing babies is frightening because the technology used would undermine the distinction between our createdness and our human freedom. As a concluding remark on this aspect of the cultural-philosophical debate, it is ironic to note, as observed by Audrey Chapman, that despite the religious undertone of the concept of "playing God," theologians involved in the ethical debate seem less inclined to use the phrase or to refer to the potential for cloning to impinge on divine prerogatives and to exceed appropriate limitations of human activity than are some secular commentators.[52]

Moralizing Human Nature[53]

Due to spectacular advances in stem cell research, more of our "natural" humanity is falling into the molding hands of biotechnology. This technological control of human nature could at first be considered another manifestation of our tendency to keep extending the range of the natural elements we can dispose of in the environment. On closer scrutiny, however, biotechnology, by enabling us to cross the line between our *outer* and *inner* nature, has led human beings to self-create artificial barriers between what is and ought not to be done. Science and technology have, for a long time, enlarged the scope of human freedom at the expense of a disenchantment with nature. This tendency has now come to an end, by erecting artificial barriers in terms of taboos, that is, by a re-enchantment of our inner nature. "Co-player of evolution" or "playing God" are the archaic remnants of expressing our revulsion at the idea of transforming the human species by cloning human beings or by manipulating embryos in the course of experimentation. In this we may see an attempt at moralizing human nature: "What science has put at our disposal on a technological level is now through moral control, in a normative perspective, to be restored to its character of something we may not dispose of."[54]

Biotechnology can be described as *symbiotic* technology[55] because it blurs the distinction between our inner and outer nature. Technology has entered our bodies in two different ways. On the one hand, by reducing its tools to microscopic sizes, it can collaborate with our organs through life-support machinery; on the other hand through genetics and genomics, the human genome can be modified and enhanced, and thus lead to an extensive transformation of human nature. Furthermore, the line between the fixed natural structures we cannot dispose of and the exercise of human freedom is further shifting when our societal structures grant researchers legal warrants to dispose of the physical being we are by nature—for example, the patentability of stem cell research.[56]

Consequently, biotechnology has been accused of incorporating human identity—perceived as genetic identity. Human identity has become, in the eyes of the opponents of ES cell research and therapeutic cloning, merely an object of choice. *Dasein* has become design. In this perspective, moralizing or even sanctifying human inner nature is a contemporary attempt, in a rigid modern environment where religious thinking, beliefs, or guarantees are lacking, to set the limits of the acceptable. Provoked by scenarios that currently, stepping out of science fiction literature, invade the scientific papers, opponents of biotechnology in general and cloning or stem cell research in particular rely on morals or mythical images to condemn its applications. By using vague mythical images, however, such as "playing God," they keep alive what D. Nelkin and M. S. Lindee called "the DNA mystique." By this religiously inspired notion they refer to the media-covered process of "geneticization": "an ongoing process by which differences between individuals are reduced to their DNA codes."[57] We would suggest that it would be better to speak of "Gene Utopia" than of "DNA mystique," in order to avoid any religious connotation.

Gene Utopia

In agreement with the Dutch philosopher Hans Achterhuis,[58] we do not want to define utopia as the German-American theologian Paul Tillich did, namely: "Being human means having utopia (*Menschsein heisst, Utopie haben*)."[59] Achterhuis wants to mark off utopia from any ambition of transcendence, from any manifestation of *Das Prinzip Hoffnung.*[60] The notion of "utopia" first appeared in 1516. Its spiritual father, Thomas More, defines utopia as a place where living is good (*ευ*) but at the same time a place that is nowhere (*ου*) to be found. The main characteristic of utopia is control: society as well as the inner and outer nature of human beings can be dominated in a rational way. More's utopia is distinct from mythical and religious paradises in light of the exclusive priority of human control. The prototype of such a technical utopia is Bacon's *New Atlantis* (1626). Bacon refers in his introduction to the story of creation. However, his world is not created by God but by human technology. Bacon explicitly denies the religious aspect of nature as being of divine creation. Nature, also human nature, lies open for scientific research and technical use. Medical technology forms a major part of it. The so-called miracles of nature at the service of the inhabitants of *New Atlantis* are prolongation of life, rejuvenation, slowing down the aging process, curing lethal diseases, and relieving suffering. To attempt an analogy, the wonder drug of *New Atlantis* could be compared to the current promises of stem cell research and therapeutic cloning.

Ted Peters, on the other hand, refers to the myth of Prometheus's pride or hubris, instead of New Atlantis, as one of the most powerful myths that frames and filters cultural reception of cloning technology.[61] This utopian consciousness, however, only arose in the sixteenth and seventeenth centuries with the rediscovery of the Greek sources of our culture, combined with a scientific revolution. Thus, we may call Prometheus's myth, described in Plato's *Protagoras,* as a proto-utopia,[62] and moreover, a Gene Utopia.

The pejorative undertone of the phrase "playing God" leads us back to our mythical past. The Greek Titan is not only characterized as creator of all sciences but also as the divine power that created humanity. The point of the story is that Prometheus provided humanity with the heavenly fire—scientific knowledge—that separates humankind from animals and promotes the human to master of the universe. From Zeus's point of view this was a transgression, which had to be followed by a fall. Zeus therefore chained the Titan to a rock where an eagle could feast all day long on his liver.[63]

Nowadays, cell powers are slapping the cuffs on humanity. On the one hand, diseases like cancer and Parkinson's make people shiver. On the other hand, these diseases can only be overcome thanks to the increasing knowledge of the therapeutic value of embryonic stem cells. According to Ted Peters, such an ambiguity may lead to a deterministic interpretation of human destiny, which he calls "Promethean Determinism." Such a cultural philosophical interpretation reduces humanity and human biology to their genetic substratum, which hardly allows human freedom and creativity to flourish. A theological clarification may therefore be helpful.

A Theological Interpretation

According to James Walter,[64] two different theological models of God and creation have been historically used in the Christian tradition to contextualize the issues that result from human biotechnological applications and appropriately address them: the stewardship model and the created co-creator model.

The Traditional Stewardship Model: Man as Steward of God's Creation

The stewardship model bears a strong resemblance to the myth of Prometheus. According to the tenants of this thesis, God is considered to be the creator of the material universe and humanity as well as the one who elaborated the fixed and universal laws of creation. God's purposes for humanity can be understood by scrutinizing the universal laws governing nature. As a sovereign ruler over the created order, God decides the future by divine providence. As a lord over life and death, God has certain rights over creation that humans do not enjoy. When humans take God's will into their own hands, they therefore usurp divine authority and thus "play God." As an anthropological counterpart, humanity is defined as the steward of God's creation. His moral responsibility primarily consists of protecting and maintaining the divine creation and order. Thus being only trustees of both their outer and inner nature (e.g., genetic heritage), humans are not delegated the moral responsibility to alter God's original creative will.

As stem cell research implies an intervention in the inner nature of humans, a theological evaluation of both the physical nature and bodily existence also plays a role in formulating a moral theological judgment on stem cell research. According to Walter, the view of physical nature that is linked with the stewardship model is sacral-symbiotic.[65] From this perspective, all materials, outer as well as inner, are considered to be created by God and thus sacred. Human biological nature is thus static and its inherent laws must be respected. Biological nature teaches humans how to live within the boundaries established by God at the moment of creation.

In other words, as noted by J.-P. Wils,[66] the stewardship model can be qualified by the following substantives: originality, exhaustiveness, stability, and optimum. *Originality* relates to the belief that nature was created from nothing, human beings included. The existence of every creature is dependent on an original initiative of someone else. This total dependence from a creator's will implies that the creating authority is omnipotent. Nature is the *exhaustive* demonstration of God's omnipotence. There is no alternative for the inner and outer nature we depend on. God's organization of nature guarantees its *stability*. Because the order of nature is original, God's creation will also be *optimal*. Creation is originally good and the goodness of creation must merely be preserved.

Adopting this theological position leads to the conclusion that ES cell research, even for the therapeutic purpose of curing lethal diseases, should be condemned. Such research constitutes an expression of human arrogance; manipulating the divine, humans take on a role that has not been delegated to them as stewards. ES cell research violates the static and inherent laws of nature—human inner nature being sacrosanct and inviolable, humans disobey the laws of nature, crossing divine boundaries.

In an age of biological control, the stewardship model may be out of date. Human knowledge gained by the biotechnological trio genetic engineering, genomics, and cloning teaches us that nature is evolving. Even "fixed" biological laws are under review, essentially because of stem cell research, for example, the capacity of adult cells to rejuvenate and to de-differentiate. Change and development are considered to be normative. "The relation between nature and human freedom appears as a dialogue that dynamically evolves over time. It is within that dialogue that humans learn how to use responsibly material reality as the medium of their own creative self-expression."[67] In addition to the fact that the stewardship model may be outdated, it runs the risk of holding a deterministic view on human nature and biotechnology—an approach that could be compared with the above-mentioned Promethean determinism.

The first model, though indicative of essential elements of theological thinking, is unsatisfactory when applied to find guidance with respect to stem cell research. The stewardship model conflicts with contemporary ideas as reshaped by the evolution of our scientific knowledge and contains the seed for a deterministic and predestined analysis of human nature.[68] In order to reconcile the exposed theory with science and draw a sound, viable, unambiguous conclusion on stem cell research, some further analysis on the concept of creation and the omnipotence of God must be conducted.

Rethinking the Traditional Theological Interpretation: The Human as Created Co-Creator

The created co-creator offers us an interesting perspective on the relationship between God and humanity, a relationship based on respect, common will, and interactivity. If when applying its precepts to stem cell research, the conclusion drawn might, at first, appear irreconcilable, then complementarity is to be found by abandoning the Promethean myth for a conception of human as true *imago Dei.*

The Notion of the Human as Created Co-Creator

When interpreting the phrase "playing God," the tension between the biblical calling of humanity to realize its essence as the image of God, on the one hand, and on the other hand its Promethean propensity to become God needs to be recalled. Human temptation to become his own creator, the one he owes his life to and worships, is at the nucleus of the doctrine of original sin as described in Genesis 3. The tension between vocation and desire may be translated as a warning against human pride: "playing God" (*imago Dei*) is not the same as "being divine" (*imitatio Dei*).[69] The expression "playing God" therefore must not be explained in terms of the human as creator but as a *created co-creator.* This term offers us a fitting key concept for a theological approach to our theme.

Whereas the Protestant theologian Philip Hefner introduced the term "created co-creator" in his article "The Evolution of the Created Co-Creator,"[70] the Catholic theologian Karl Rahner S.J. had, in the early seventies, anticipated a similar vision of humanity in relation to God and creation.[71] We now propose unraveling the several components of this expression and evaluating them one by one.

The term co-creator implies that creation or nature is not a static order brought into existence through a single, one-off, act of God, but rather a work in progress. This is evidenced by our ever-growing scientific knowledge, especially in the field of genetics. Revolutionary genetic developments have put theological anthropology into an evolutionary perspective. The DNA molecule is shared by all living beings. Spontaneous genetic mutations have not only led to human evolution, but have been and are also responsible for faults in the mechanism of cell division that often lead to fatal illnesses such as cancer. From the perspective of God the new insights gained from genetics mean that creation is an evolutionary process in which God remains continuously active. However, God remains ubiquitous and can at every moment influence creation at the level of the smallest building blocks (the hereditary material). In the end, the future of this evolution is always uncertain, just as God's kingdom is found in the field of tension between the "already" and the "not yet." These elements constitute the kernel of the doctrine of the uninterrupted creation, or "*creatio continua*," of God.[72] For humanity, knowledge of the origins of a (genetic) disease means that the human is no longer a passive subordinate of God's creation but may take an active and creative part in it through, for example, the recourse to (therapeutic) cloning technology.

The prefix "co" in "co-creator" suggests that the human stands on an equal footing with God. Thanks to the development of ES cell research, humans have indeed gained insight into one of the secrets of God's creation. With the aid of cloning techniques they are able to apply this knowledge to the creation of human tissues (therapeutic cloning). But this does not necessarily imply that the microbiologist will always use this knowledge for the greater good of humanity, the betterment of God's creation (e.g., reproductive cloning).[73] In opposition to God's intentions, which are inherently good, human creativity is ambiguous. Scientists may apply their technological ingenuity to reduce human suffering, but may also cause such suffering. An unbounded veneration for the efficiency, measurability, and usefulness of human life has precisely been the cause for the exploitation of all creatures—including (beginning with) human life—as objects, means, or functions of humans themselves. The human is not divine; the human only plays at being God and is only an image of God. "When man loses perspective of the ambiguity of his creativity, the risk is great that he confuses the technologically 'possible' with the 'desirable' or the 'valuable,' in short, with the 'good.'"[74]

To the notion of co-creator the complement "created" must necessarily be added. God's work of creation is always distinct from that of humans. While the human being only disposes of the natural properties of already available matter, God creates from nothing (*creatio ex nihilo*). Nature in its entirety is dependent on the divine creator who transcends it. If this distinction between the divine and human creation is not respected "it betrays a veiled naturalism, a variant on the alleged 'thou shalt not play God' commandment."[75] It supposes that human life when naturally gifted (in whatever way) has a higher moral value than nature altered by scientific intervention, supported by technology. Applied to ES cell research, this would mean that the stem cell is a natural "biological niche" for the unique genetic constitution of (beginning) human life that may not be humanely altered without violating human dignity. The fact that stem cells are researched and manipulated by individuals would in itself make cloning even for therapeutic ends immoral: "what Mother Nature has given us is inviolable."

Such an argument contains a naturalistic fallacy. It presumes to deduct what *ought* to be from what *is*. But Mother Nature is not divine. God has created nature as good, but the creation is not complete (cf. *creatio continua*): suffering, sickness, and death are equally the consequences of our genetic constitution. Human cloning technology could thus aim at the improvement of our natural (divine) gift inherited from creation. The human hereditary material itself is not holy—only God is. Because life is created, it is good; but it is not God.

Applying the precepts of this model to form a theological opinion on the use of the new cloning technologies for therapeutic purposes, it seems necessary to constantly bear in mind two essential consequences of the "created co-creator" concept. On the one hand, the very nature of humankind, created by God, is good. Although God has created humans in his image (naturally good), his work is still in progress, mishaps occurring in the form of diseases, suffering, and death. Humanity, in the image of God, is therefore called to apply technology to work with God in the unending kingdom (cf. the created *co-creator*) and improve the sketch of creation. On the other hand, the human is but a poor and imperfect image of God; he is a created image who does not enjoy the same status as the model. Technological intervention when performed by men is subject to their sins (cf. the *created* co-creator). Separating these components of the human being as created co-creator can lead to two one-sided and irreconcilable judgments of the merits of ES cell research.

First, if we understand the human being only as a co-creator, humans should consider (human) life as valuable only in terms of its usefulness for humanity itself. Human embryonic stem cells could thus be regarded as raw material to be manipulated in function of human needs or desires. In that regard, ES cell research should be allowed. One small specification needs to be made, however. Scientifically envisaged, embryonic stem cells cannot be considered as equivalent to human life, and microbiologists are, for the moment, "borrowing" them from human embryos in vitro. Apart from and independent from the evaluation of the potential advantages of ES cell research, scientists must always keep in mind that "weak" (beginning) human life is also created in the image of God and therefore deserves respect.

Second, if we regard human beings only as created beings alongside other creatures, subordinates of God, they cannot be allowed to intervene in matters of human life. Thus, it becomes difficult to justify not only cloning technology but also medicine in general, for all of medical science, and not only ES cell research, aims at the reduction of human suffering. Nevertheless, the notion of "createdness"—and with it mortality and fallibility—in combination with the doctrine of sin provides humanity with the important insight that not everything that can be done may be done.

Imago Dei versus Promethean Hubris

The key question isn't *whether* we play God but rather *how* we should play God. Knowing good and evil and how to differentiate them, humanity has to realize and incarnate his or her morally good disposition, following Jesus Christ, who is the true image of God. The Christian conception of God is inseparable from the stories of Jesus of Nazareth, the Son of God, whom Christians believe to be the definitive expression of God's purpose for humanity. Although the four evangelists offer different portraits of Jesus' life, they all agree on one point: Jesus of Nazareth spent a great deal of his life healing the sick. If the Gospels present Jesus' healing miracles as an essential element of his divine identity, does this then mean that he defies the original will of the creator? Or are the miraculous healings that Jesus accomplished the very expression of God's intention for humans?

If Jesus were thus to have defied the will of God by healing the sick, would he be placed on the same level as Prometheus, who rebelled against the tyranny of the evil chief divinity Zeus? Jesus himself fulminated against such a vision when he said, "I drive out demons with the help of the spirit of God"[76] (Matt. 12:22–32). Jesus, moreover, on three occasions, withstood the invitation of Satan to abuse his divine powers and collaborate in the rebellion against God (Matt. 8:1–10). Jesus' miraculous healings perfectly coincide with the conception of the human as image of God (Exod. 15:26b), while Eden remains the symbol of the original and final aim of God's work of salvation (Isa 51:3).[77] The healing miracles of the New Testament confirm the Old Testament's portrayal of a God who takes pleasure in the (human) nature of creation and at the same time continuously creates and redeems (human) nature, a process culminating in the life, death, and resurrection of Christ. According to Christians, (human) nature—marked by sickness, suffering, and death—is not consistent with God's original creative work. (Human) nature is originally good but at the same time is subjected to evil forces due to human freedom with regard to sin. Rudolf Bultmann sums up the ambiguous character of nature as follows: "The creation has a peculiarly ambiguous character: on the one hand, it is the earth placed by God at man's disposal for its use and benefit . . . ; on the other, it is the field of activity for evil, demonic powers."[78]

In sum, we can say that emphasizing the "co-creator" pole of humankind as *imago Dei* implies an invitation to accept the advantages of stem cell research. Purely stressing the fact that the human is created by God, but sinful and therefore fallible, may lead to a condemnation of cloning techniques merely with regard to certain purposes. In the concluding section, the acceptance of human fallibility will be the basic assumption for a moral-theological evaluation of stem cell research.

Conclusion: A Moral-Theological Evaluation

The ethical discussion surrounding stem cell research is trying to answer questions such as "what can we do?" "what may we do?" and "what do we have to do?" to enlarge our physical capabilities. These questions concern the permissibility of our biotechnological applications that aim at enhancing our natural condition, our physical and psychological well-being.

The theological interpretation, by insisting on human fallibility and its sinful tendencies, introduces *passivity* into the ethical discussion. This concept suggests that because of human imperfection and sense of morality, we, as humanity, will either be faced with the limits of our technological capabilities or refuse to perform certain interventions on moral grounds. Introducing this notion of passivity into the clinical or research setting might, however, prove a sensitive and delicate task. Whoever casts doubt on the acceptability of a certain intervention or on the relevance and urgency of a particular scientific research, such as stem cell research, will probably have to cope with the following question: "Assume that you or one of your relatives would have a lethal disease, and that there may be an opportunity to overcome that disease, wouldn't you promote stem cell research?"[79]

Medical, biotechnological, or scientific advances and research tend to be considered subjectively in terms of achievement and progress and of the power they have provided humans in matters of health and life. However, one needs to equally notice humanity's limitations and inability to rid the earth of illnesses, suffering, and death. Although progress needs to be welcomed, it is equally imperative to see the world as it is and to learn to live and deal with its incapacities, which are inherently connected to the human condition. From this perspective, the ethics of stem cell research has to do not only with feasibility or activity but also with passivity. This ethics of passivity is an active exercise, namely, the capacity to be aware of human (bio)technological limits. In short, there are things we cannot solve, but only have to cope with. In the words of T. S. Elliot: "There are two kinds of problems in life. One kind requires the question 'What are we going to do about it?,' and the other calls forth different questions: 'What does it mean? How does one relate to it?'"[80]

To clear up any misunderstanding, we promote attempts to reduce suffering, illness, and premature death by technical means. However, at this point in human history, these realities (suffering, illness, etc.) cannot be ignored, and Gene Utopia is only used to evade reality. Although a longing for invulnerability is perhaps a quintessential human trait and the quest to reduce human suffering wrought by illness and disease is morally admirable, there is no mistaking the hubris behind the question "Why die?" Therefore we agree with Paul Lauritzen that we need to ask whether we wish to accept and promote a view of bodily vulnerability as merely an obstacle to human flourishing, which ought to be overcome at any cost.[81]

Many lethal diseases are still to be symbolically accepted as a part of our *condition humaine*. Utopial visions of controlling our genetic destiny and history by means of stem cell transplantation are just contributing to increasing the pain of a lot of patients. To illustrate this, we can refer to the growing European consensus that ES cell research can be carried out by using "spare" embryos. This approach seems to imply that such research is not controversial. However, the use of spare embryos requires the creation of such embryos by IVF technique. IVF procedures' success rate (i.e., a pregnancy) stands no higher than 20 percent. It is in fact a treatment in which the recurrence of suffering and disappointment is a "normal" side effect. We may even say that "IVF is as unpredictable as a stork."[82] In order to undertake a sound ethical evaluation of stem cell research, its potential as well as its risks, shortcomings, and limits need to be taken into consideration and balanced. Evaluating stem cell research is like throwing a pebble into the water. It makes expanding rings on the water's surface. In the center of the circular movement is the embryo. A second ring includes the parents who have decided to donate their embryo for scientific research. A third one includes the "early patient" who needs to be treated, and finally there is the circle of solidarity among human beings.

Because of the historicity of every human person, the theological notion of sin or its ethical counterpart passivity suggests that we should constantly reconsider what current possibilities would better promote the human person. The ethics we propose is dynamic in that it reevaluates its findings according to the evolution of our scientific knowledge. As science and technology constantly advance, new possibilities can be shaped by our creativity, for example, therapeutic cloning. Ethics has to inquire how the growing possibilities can be made use of to better serve the dignity of humankind. The promotion of the human person becomes a moral obligation insofar as it is possible (*le souhaitable humain possible*).[83] The task of ethics, as a way of thinking and reflecting on life, consists in keeping step with human life itself as it unfolds throughout history.

A proportionate ethical and moral weighing of values and disvalues guides us and enables us to overcome the reduction of human freedom to the right to "be left alone" or to the right to "dispose over our human life and body," that is, to dispose over our human genome. Keeping the humanly desirable in perspective, humankind is called to create a world inhabited by human dignity and to avoid the reduction of human beings to purely instrumental entities. This represents a continuous, but very creative challenge: as a *created co-creator,* every human being may assume his or her full responsibilities to take creation in line with God's intentions one step further toward completion.

Notes

1. G. de Wert, *Het schaap kan nu worden gekopieerd, de volgende stap is de gekloonde herder* [A sheep can be copied, the next step will be a cloned shepherd], in *NRC Handelsblad* (March 12, 1997): 7.

2. R. Cole-Turner, "The Era of Biological Control," in *Beyond Cloning: Religion and the Remaking of Humanity*, ed. R. Cole-Turner (Harrisburg, Pa.: Trinity Press International, 2001), 1–2. Cloning and stem cell technology lay the foundation of what Ian Wilmut, Dolly's godfather, calls the "age of biological control." See I. Wilmut, K. Campbell, and C. Tudge, *The Second Creation: Dolly and the Age of Biological Control* (New York: Farrar, Straus & Giroux, 2000).

3. "Human Cloning Needs a Moratorium, Not a Ban," *Nature* 386 (1997): 1.

4. K. Hochedlinger and R. Jaenisch, "Nuclear Transplantation, Embryonic Stem Cells, and the Potential for Cell Therapy," *New England Journal of Medicine* 349 (2003): 275–86.

5. P. Verspieren, "Le clonage humain et ses avatars," *Etudes* 391 (1999): 459–67: "Ce qui a suscité jusqu'à présent le plus de débats est l'application à l'homme de ce qui est communément appelé 'clonage reproductif'; il vaudrait mieux parler de clonage *à visée reproductif*."

6. B. Vogelstein, B. Alberts, and K. Shine, "Please Don't Call It Cloning," *Science* 295 (2002): 1237–38.

7. P. R. McHugh, "Zygote and 'Clonote'—The Ethical Use of Embryonic Stem Cells," *New England Journal of Medicine* 351 (2004): 209–11.

8. Department of Health, *Stem Cell Research: Medical Progress with Responsibility* (London, 2000), 27–28 (http://www.dh.gov.uk/assetRoot/04/06/50/85/04065085.pdf [access February 10, 2005]).

9. Bioethics Advisory Committee, *Advies nr. 10 van 14 juni 1999 i.v.m. het reproductieve menselijke klonen* [Advice no. 10 of June 14, 1999 concerning human reproductive cloning] (Brussels, 1999), 8.

10. Nuffield Council on Bioethics, *Stem Cell Therapy: The Ethical Issues* (1999) 3–4 nn. 3 and 6 (http://www.nuffieldbioethics.org/go/ourwork/stemcells/introduction [access February 10, 2005]).

11. J. A. Robertson, "Ethics and Policy in Embryonic Stem Cell Research," *Kennedy Institute of Ethics Journal* 9 (1999): 110.

12. Council of Europe, *Convention for the Protection of Human Rights and Dignity of the Human Being with Regard to the Application of Biology and Medicine. Convention on Human Rights and Biomedicine* (Oviedo, 1997), 6 (http://conventions.coe.int/Treaty/en/Treaties/Html/164.htm [access February 10, 2005]).

13. D. Dickson, "Ethics Can Boost Science," *Nature* 408 (2000): 275. It has to be noticed that because of the therapeutic value of ES cell research,

France and the Netherlands signed the convention but did not ratify it. They foresee making an exception for article 18.

14. D. Josefson, "Adult Stem Cells May Be Redefinable," *British Medical Journal* 318 (1999): 282.

15. G. Vogel, "Harnessing the Power of Stem Cells," *Science* 283 (1999): 1432–34.

16. Pontifical Academy for Life, *Declaration on the Production and the Scientific and Therapeutic Use of Human Embryonic Stem Cells* (Vatican City, August 25, 2000): "The progress and results obtained in the field of adult stem cells show not only their great plasticity but also their many possible uses, in all likelihood no different from those of embryonic stem cells, since plasticity depends in large part upon genetic information, which can be reprogrammed."

17. S. Pagán, "Ultimate Stem Cell Discovered," *New Scientist* (January 23, 2002). Recently, however, two teams of scientists have published research that suggests that adult stem cells may not have as much potential to develop into other kinds of body cells as was thought. See G. Vogel, "Stem Cell Research. Studies Cast Doubt on Plasticity of Adult Cells," *Science* 295 (2002): 1989.

18. UNESCO, *Universal Declaration on the Human Genome and Human Rights* (Paris, 1997) http://www.unesco.org/shs/human_rights/hrbc.htm (accessed July 17, 2004).

19. Ibid., art. 11.

20. European Parliament, *Resolution on Human Cloning* (January 15, 1998) OJ C 34, p. 164; http://www.europarl.eu.int/comparl/tempcom/genetics/links/b4_050_en.pdf (accessed July 17, 2004).

21. European Parliament, *Resolution on the Cloning of the Human Embryo* (October 28, 1993) OJ C 315, p. 224; http://www.europarl.eu.int/comparl/tempcom/genetics/links/b3_1519_en.pdf (accessed July 17, 2004); idem, *Resolution on Cloning Animals and Human Beings* (March 12, 1997) OJ C 115, p. 92; http://www.europarl.eu.int/comparl/tempcom/genetics/links/b4_ 0209_en.pdf (accessed July 17, 2004); idem, *Resolution on Human Cloning* (September 7, 2000) http://www.europarl.eu.int/comparl/tempcom/genetics/ links/b5_0710_en.pdf (accessed July, 17, 2004).

22. Council of Europe, *Additional Protocol to the Convention for the Protection of Human Rights and Dignity of the Human Being with Regard to the Application of Biology and Medicine: On the Prohibition of Cloning Human Beings*, (Paris, 1998) http://conventions.coe.int/Treaty/en/Treaties/Word/168. doc (accessed July 17, 2004).

23. Comité Consultatif de Bioéthique de Belgique, *Avis n° 10 du 14 juin 1999 concernant le clonage reproductif* (Brussels, 1999) http://www.health. fgov.be/bioeth/fr/avis/avis-n10.htm (accessed July 17, 2004). An English version of the opinion is available upon request at the following address: bioeth-info@health.fgov.be.

24. United Nations, *Resolution 56/93 International Convention on the Cloning of Human Beings* (2001) http://ods-dds-ny.un.org/doc/UNDOC/GEN/N01/479/51/PDF/N0147951.pdf?OpenElement.

25. United Nations, *Annex I to the Letter Dated 2 April 2003 from the Permanent Representative of Costa Rica to the Untied Nations Addressed to the Secretary-General. Draft International Convention on the Prohibition of All Forms of Human Cloning* (2003). http://ods-dds-ny.un.org/doc/UNDOC/ GEN/N03/330/84/PDF/N0333084.pdf?OpenElement.

26. For more information on the work of the United Nations, see http:// www.un.org/law/ cloning/; see also the Center for Genetics and Society Web site: http://genetics-and-society.org/policies/international/un.html.

27. Conference videos and papers can be consulted at http://europa.eu.int/comm/research/quality-of-life/stemcells.html.

28. European Group on Ethics in Science and New Technologies, *Adoption of an Opinion on Ethical Aspects of Human Stem Cell Research and Use* (November 14, 2000; rev. ed., January 2001).

29. House of Lords, *Stem Cell Research: Report* (February 13, 2002) http://www.parliament.the-stationery-office.co.uk.

30. http://www.bionews.org.uk/new.lasso?storyid=2096.

31. Wet van 11 mei 2003 betreffende het onderzoek op embryo's in vitro, *Belgisch Staatsblad* (May 28, 2003); Loi de 11 Mai 2003 relative à la recherche sur les embryons in vitro, *Moniteur Belge* (May 28, 2003). http://www. staatsblad.be.

32. Sénat, *Projet de Loi Relatif à la Bioéthique (Texte definitif)* (Paris, July 8, 2004) http://www.senat.fr/pl/106-1ere-partie-0304.pdf. It is however to be noted that some deputies have submitted the final text to the "Conseil Constitutionel" in order for its conformity to the constitution to be checked. As a result the new act is not yet binding law.

33. Wet van 20 juni 2002, houdende regels inzake handelingen met geslachtscellen en embryo's (http://www.wetten.overheid.nl).

34. Gesetz vom 28. Juni 2002 zur Sicherstellung des Embryonenschutzes im Zusammenhang mit Einfuhr und Verwendung menschlicher embryonaler Stammzellen, *Bundesgestzblatt* 29 (June 2002). http://217.160.60.235/BGBL/ bgbl1f/BGBl102042s2277.pdf.

35. See http://escr.nih.gov/. On August 9, 2001, at 9:00 p.m., President George W. Bush announced his decision to allow federal funds to be used for research on existing human embryonic stem cell lines as long as prior to his announcement (1) the derivation process (which commences with the removal of the inner cell mass from the blastocyst) had already been initiated and (2) the embryo from which the stem cell line was derived no longer had the possibility of development as a human being. The National Institutes of Health, proceeding under this compromise, has since made fifteen to twenty human stem cell lines available for federally funded research.

36. For a more extensive review see, e.g., C. M. Romeo-Casabona,

"Embryonic Stem Cell Research and Therapy: The Need for a Common European Legal Framework," *Bioethics* 16 (2002): 557–67.

37. P. Verspieren, "Le clonage humain."

38. K. Rahner, "The Problem of Genetic Manipulation," *Theological Investigations* 9 (1972): 246.

39. R. A. McCormick, *The Critical Calling* (Washington, D.C.: Georgetown University Press, 1989), 343–46.

40. M. A. Farley, "Roman Catholic Views on Research Involving Human Embryonic Stem Cells," in *The Human Embryonic Stem Cell Debate: Science, Ethics and Public Policy*, eds. S. Holland, K. Lebacqz, and L. Zoloth (Cambridge, Mass.: MIT Press, 2001), 117. The author also refers to T. A. Shannon and A. B. Walter, "Reflections on the Moral Status of the Preembryo," *Theological Studies* 51 (1990): 603–26; and L. S. Cahill, "The Embryo and the Fetus: New Moral Contexts," *Theological Studies* 54 (1993): 124–42, who do not consider the human embryo in its earliest stages to constitute an individualized human entity with the settled inherent potential to become a human being.

41. Cf. G. Outka, "The Ethics of Human Stem Cell Research," *Kennedy Institute of Ethics Journal* 12 (2002): 175–213.

42. C. Tauer, "Responsibility and Regulation: Reproductive Technologies, Cloning and Embryo Research," in *Cloning and the Future of Human Embryo Research*, ed. P. Lauritzen (New York: Oxford University Press, 2001), 153.

43. C. Holden, "Stem Cell Research: Primate Parthenotes Yield Stem Cells," *Science* 295 (2002): 779–80.

44. E. Dorff, "Resolution in Support of Stem Cell Research and Education" (www.rabassembly.org), cited in P. Lauritzen, "The Ethics of Stem Cell Research," in *Monitoring Stem Cell Research,* by the President's Council on Bioethics (www.bioethics.gov).

45. A. R. Chapman, *Unprecedented Choices: Religious Ethics at the Frontiers of Genetic Science* (Minneapolis: Fortress Press, 1999), 52–57.

46. T. Peters, *Playing God: Genetic Determinism and Human Freedom* (London: Routledge, 1997) 2, 10–11.

47. R. Dworkin, *Sovereign Virtue: The Theory and Practice of Equality* (London: Harvard University Press, 2000), 427–52.

48. S. Graumann, "Experts for Philosophical Reflection in the Public Discourse: The German Sloterdijk-Debate as an Example." Unpublished lecture at the Euresco Conference on Biomedicine within the Limits of Human Existence, September 8–13, 2001, Davos, Switzerland.

49. P. Sloterdijk, "Regeln für den Menschenpark," *Die Zeit* (September 16, 1999): 17–21.

50. A. Vollmer, "Ridders van de Hypermoraal [Knights of hyper moralism]," in *Regels voor het mensenpark: Kroniek van een debat* [Rules for the human breeding ground: chronicle of a debate], ed. P. Sloterdijk, (Boom/Amsterdam, 2000), 102–7.

51. J. Habermas, *Die Zukunft der menschlichen Natur: Auf dem Weg zu einer liberalen Eugenik?* (Frankfurt: Suhrkamp, 2001), 80–84.

52. A. R. Chapman, *Unprecedented Choices,* 93.

53. Throughout this paragraph, we follow Jürgen Habermas's line of reasoning (*Die Zukunft der Menschlichen Natur*).

54. J. Habermas, *Die Zukunft der Menschlichen Natur,* 46: "Was durch Wissenschaft technisch disponibel geworden ist, soll durch moralische Kontrolle normativ wieder unverfügbar gemacht werden."

55. J.-P. Wils, "Body, Perception and Identity." Unpublished lecture at the Euresco Conference on Biomedicine within the Limits of Human Existence.

56. G. Van Overwalle, personal communication: "A key example of a patent which has been granted for human ES cell research is U.S. patent 6.200.806, entitled 'primate embryonic stem cells.' The patent related to an invention from James Thomson from Madison and was delivered to the Wisconsin University of Madison on March 13, 2001."

57. A. Lippman, "Prenatal Genetic Testing and Screening: Constructing Needs and Reinforcing Inequities," *American Journal of Law and Medicine* 17 (1991): 19; cited in K. Dierickx, *Genetisch gezond? Ethische en sociale aspecten van genetische tests en screenings* [Genetically healthy? Ethical and social aspects of genetic testing and screening] (Antwerp, 1999), 177 n. 204.

58. H. Achterhuis, *De erfenis van de Utopie* [utopia's heritage] (Amsterdam, 1998), 14–20.

59. M. J. Laski, *Utopia and Revolution* (Chicago: University of Chicago Press, 1976), 641; cited in H. Achterhuis, *De erfenis van de Utopie,* 14.

60. E. Bloch, *Das Prinzip Hoffnung* (Frankfurt: Suhrkamp, 1959); cited in Achterhuis, *De erfenis van de Utopie,* 14.

61. E.g., J. R. Meyer, "Human Embryonic Stem Cells and Respect for Life," *Journal of Medical Ethics* 26 (2000): 166–70.

62. H. Achterhuis, *De erfenis van de Utopie,* 179.

63. S. Golowin, M. Eliade, and J. Campbell, *De grote mythen van de wereld* [Myths of the world] (Leuven, 1999), 72–74.

64. J. J. Walter, "Human Gene Transfer: Some Theological Contributions to the Ethical Debate," *Linacre Quarterly* 68 (2001): 319–34, esp. 324–27. See T. A. Shannon and J. J. Walter, *The New Genetic Medicine: Theological and Ethical Reflections* (Lanham, Md.: Rowman & Littlefield, 2003).

65. J. J. Walter, "Human Gene Transfer," 326.

66. J.-P. Wils, "Body, Perception and Identity."

67. J. J. Walter, "Human Gene Transfer," 326–27.

68. In this connection it may also be instructive to return to the distinction between the sacred and the holy, as Roger Burggraeve reflectively develops it. See R. Burggraeve, "Biblical Thinking as the Wisdom of Love,"

in *Anti-Judaism and the Fourth Gospel*, ed. R. Bieringer, D. Pollefeyt, and F. Vandecasteele-Vanneuville (Louisville, Ky.: Westminster John Knox Press, 2001), 202–25.

69. E. Schroten, "Some Theological Comments on a Metaphor," in *Christian Faith and Philosophical Theology*, ed. G. Van de Brink et al. (Kampen: Kok Pharos, 1992), 192–94.

70. P. Hefner, "The Evolution of the Created Co-Creator," in *Cosmos and Creation: Theology and Science in Consonance*, ed. T. Peters (Nashville: Abingdon, 1989), 211–33. See also J. J. Walter, "Human Gene Transfer," 332 n. 25: "James Gustafson *(Ethics from a Theocentric Perspective, Vol. II.*, Chicago, 1984, p. 294) has preferred to describe our role in creation as 'participants' rather than as 'co-creators.' He argues that the divine continues to order creation, and we can gain some insight into God's purposes by discovering these ordering processes in nature."

71. K. Rahner, "The Experiment with Man: Theological Observations on Man's Self-Manipulation," *Theological Investigations* 9 (1972): 205–24; idem, "The Problem of Genetic Manipulation," *Theological Investigations* 9 (1972): 225–49.

72. R. Cole-Turner, *The New Genesis: Theology and the Genetic Revolution* (Louisville, Ky.: Westminster John Knox Press, 1993), 98–99.

73. Ibid., 102–3.

74. R. Burggraeve, *De Bijbel geeft te denken* [The Bible makes us reflect], 59, o.c.

75. T. Peters, "Cloning Shock: A Theological Reaction," in *Human Cloning: Religious Responses*, ed. R. Cole-Turner (Louisville, Ky.: Westminster John Knox, 1997), 20.

76. K. Dierickx, *Genetisch gezond?* 148. "The miracle-healings of Jesus can be divided in two groups. The first group consists of exorcisms, and these belong to the most authentic nucleus of Jesus' traditions. The second group of stories was shaped by the healings, more properly speaking, and deals with the curing of deaf, blind and lame people."

77. R. Cole-Turner, *The New Genesis,* 80–84.

78. R. Bultmann, *Theology of the New Testament*, 2 vols. (New York, 1951, 1955); cited in R. Cole-Turner, *The New Genesis,* 84.

79. P. van Tongeren, *Morele passiviteit* [Moral passivity], in *De nieuwe mens: Maakbaarheid in lijf en leven* [The new mankind: feasibility of body and life], ed. F. de Lange (Kampen, 2000), 20–24.

80. Ibid., 28.

81. Lauritzen, "Ethics of Stem Cell Research."

82. Achterhuis, *De erfenis van de Utopie*, 354–61.

83. P. Ricoeur, "Le problème du fondement de la morale," *Sapienza* 28 (1975): 313–17.

3

The Human Being and the Myth of Progress; or, The Possibilities and Limitations of Finite Freedom[1]

DIETMAR MIETH

Evil and the Monsters in Dystopias

The monsters of fantasy literature or horror films, such as Goethe's *homunculi,* are reduced creations fashioned by human hands and are not to be equated with evil. "Evil"—if it is at all legitimate to use this concept from metaphysics—is an abstraction, a general concept for bad and wrong actions of human beings, as differentiated from good and right actions. In a book written with Karl Rahner, *Das Böse und die Bewältigung des Bösen in Psychotherapie und Christentum* (Freiburg im Br.: Herder, 1982), 23, Albert Görres evokes Ludwig Wittgenstein to formulate this precisely:

> if there were only an accurate sentence in ethics. . . . Now there is one: I have rights. *Humans have rights, therefore there is good and evil* [because rights can be violated]. . . . When I speak of evil, I mean not more and not less than injustice. Whether there could be another concept of evil remains open [there is, but it is metaphorical: the destructive, the terrible, the cause of moral horror]. I only mean this one. Sickness, disease, suffering, fear, and death are great ills, but not every ill is necessarily evil. This is what we call the ill in the moral realm.

Monsters are figments of the imagination, fictive inventions of exaggerated yet possible human actions or human dealings (for example, King Kong). The human beings acting and creating are morally responsible. Most monsters, beginning with Frankenstein, are "creatures" of a second order, in which the human being reaches the limits of his or her creative freedom by failing to recognize these in their finiteness. Moreover, monsters are creatures in *bodily* form, even though these are dead bodies at times (e.g., vampires, Dracula). Evil, however, is a spiritual phenomenon that encompasses mental actions. Because *evil*, particularly as an abstract epitome or as an inordinately mysterious metaphor (*mysterium iniquitatis*), is spiritual, the *evil one* appears as the fallen angel (Lucifer, the "light-bringer") in Jewish–Christian–Islamic mythology. The devil is, as spirit, closer to God than the human being is. This is illustrated by Job's narrative, which Goethe adapted at the beginning of *Faust*. Monsters are mediated through the human being and his or her derived creative power; they encounter the human being as their demiurge, their God, and raise the theodicy question: "You Human-God, why have you created me like this, abandoned and without a partner? (Frankenstein's words!)

Monsters, as Kant established through recourse to natural teleology (the purpose of nature), are "purposeless." They are not fulfillments of the normal functionality of nature, for example, in a two-headed calf or in Siamese twins. Neither the calf nor the twins lose the respect due them, although quite differently, as a living thing or as a human being. Yet the human being instrumentalizes them, makes much of them, has displayed them for centuries in sideshows at fairs or circuses. Hence, the derivation of the word "monster" from the verb *monstrare*, "to show." The showpiece, that which is shown, the *monstratum*, is the monster. (In contrast, the sacred *monstrans*, for example, is that which shows itself actively, not passively.) The unusual, "purposeless," uncategorizable is placed by human beings on a pedestal (often literary or cinematic) as the "monstrous." As an exception it proves the rule, and, shuddering contentedly, we establish that we are "normal."

But what if the unusual, the abnormal, becomes normal? Taking Stanislas Lem as an example, we will show that this awaits us in literary worlds of science fiction. And what happens if our "normal" distinctions between "good" and "evil," "right" and "wrong" lose their validity? Sometimes "brave, new worlds" (Aldous Huxley) arise, in which the "artificial" replaces the "natural" culture of the world we live in. Is it permissible for us to imitate natural chance, which does not bear responsibility, and to elevate the monstrous to a new convention of a different life environment?

Sigmund Freud called evil *das Rücksichtslose,* the inconsiderate, the ruthless, the thoughtless, and this in the sense of lacking retrospection (*Zurücksicht*), of failing to remember, as well as of lacking consideration of the rights of others. Psychoanalysis, with its assumption that there is a human origin of the tendency toward evil and destruction (Erich Fromm: the tendency toward lifelessness, necrophilia), seems to offer our age more appropriate models of comprehension than behavioral research. The latter views all so-called evil as a waste product of evolution or, following the theologian Teilhard de Chardin, of an approaching noosphere, of a spiritual future.

For our purposes, we comprehend evil as a qualification of the bad and wrong acting of the human being, as his or her false creatorship, or as the misuse of his or her creative freedom. This results from the nonrecognition of human finitude, that is, lacking the insight that the human being has only limited knowledge and ability, makes mistakes, and does not have absolute control over the irreversible consequences of his or her historical acting.

Theologians such as Teresa of Avila (sixteenth century) or Paul Tillich see evil as a result of the nonrecognition of finiteness, that is, of "creatureness." At the same time, it is an expression of distancedness from God, and, as such, both a consequence and a source of human injustice. Relation to God is not a supplement altering the independence of the ethical dimension but rather an intensification of its significance.

Since the acting of human beings is subject to the unknown course of history, evil can even give rise to good. This is what Goethe envisioned in Mephistopheles's lines [Ich bin] "ein Teil von jener Kraft, die stets das Böse will und stets das Gute schafft" ([I am] part of that power that always intends evil and always creates good).[2] The irony of history! It also becomes evident in tales of the good impulses of monsters, especially toward women and children. Evil, too, is "infected" with good, as good, unfortunately, is with evil. The quotation can just as easily be reversed: The human being is "ein Teil von jener Kraft, die stets das Gute will und stets das Böse schafft" ("part of that power that always intends good and always creates evil"). This appears to be valid for many "good" impulses in the natural sciences and medicine today. Intending good: today that would be, for example, therapeutic goals and options. Creating evil—in the sense of injustice—is in turn connected with the means accepted to do so: the selection of human living things in prenatal diagnostics and their instrumentalization and utilization in research. What is future, nonexistent but at any rate possible, serves to legitimize real existing "victims." Or it is simply accepted that future risks are uncertain because we must expect a number of unknowns (e.g., in the event of interventions in the human germ line for therapeutic purposes).

The instrumentalization of living things is again an inquiry into the good intentions of the life sciences. Is life still seen as a unique, self-propelling form of being or has it not long since been seen as "biological material," as the wording reads in a 1998 guideline of the European Union allowing human life to be patented? In the book *Designing Life? Genetics, Procreation, and Ethics*, the foundations of human genetics are presented under the title "The Machine in Man."[3] Such metaphors reflect the growing consciousness of the materialization of life anticipated in the literary and cinematic monsters.

The Creative Power of the Human Being and the Normative Power of the Fictive

Must we not also fear that we could fail to do good if we delay or check "advances" in knowledge, competence, and production (for example, through moratoria)? Refraining from action is a part of action. This brings Wilhelm Busch's reductive characterization of the good to mind: "Das Gute, dieses stehet fest, / ist stets das Böse, das man läßt" (good is always the evil that one doesn't do). But what I mean here with my perspective on inaction as action is the reverse of what is prevalent today: "Das Böse, dieses stehet fest, ist stets das Gute, das man läßt" (evil is always the good that one doesn't do). And, indeed, we are responsible for that which we do as well as for that which we do not do. While some see artificial intelligence and genetic engineering as Pandora's box, a gift which shouldn't be opened, others see in the inaction a refusal to allow the potential humanization of our world. Here it is imperative to differentiate the two positions. The following thoughts on the creative power of the human being and its limitations, on the forgotten finitude of the human being, on the end of Goethe's tragedy *Faust II*, with its ironic depiction of the fundamental human situation, and on the cultural amnesia prevalent today will serve this purpose.

The creative power of the human being involves the acting of the human being; creativity would not become visible if nothing were created by the human being. Ancient Greek philosophers made a distinction between different kinds of creative activity: (1) *technique*, the skill, or as we would put it today, the know-how necessary to create something. Then (2) the *poesis*, still existent in the word "poesy" or "poetry." *Poesis* emphasized that a work had been created—today we see this in the narrower concept of poetics limited to the work of art. *Poesis* meant the creation of works, and the emphasis was placed on the result, the product, the work. Finally, (3) the *praxis* of human beings. Today we speak of *praxis* in a very broad sense, but originally *praxis* meant the acting of human beings that has an effect on the human being carrying out the action. How he or she affects himself or herself—this is the connotation of *praxis*, and this is why *praxis* essentially is also related to ethics. How the human being effects changes in himself or herself through his or her actions is a central question of ethical responsibility.

As an ethicist I am concerned with the question of *praxis*, with its (retrospective) effect on the human being: how does that which the human being creates affect himself or herself, and what is the correct model for this? To approach this problem I will briefly discuss two relevant books. The first book was written by the American David F. Noble and bears the title *The Religion of Technology: The Divinity of Man and the Spirit of Invention* (New York: Alfred A. Knopf, 1997); the second, by Bernd Gräfrath, bears the title *Es fällt nicht leicht, ein Gott zu sein* and the subtitle *Ethik für Weltenschöpfer von Leibniz bis Lem* (Munich: Beck, 1998). Stanislas Lem, a famous science fiction writer, has repeatedly thematized the question of the morality of the future.[4]

David Noble draws attention to the American dream of deliverance through technology. One of the most important writers he introduces—new to me, too—is the American sociologist Edward Bellamy (1850-1898). Bellamy represents most strongly the robust spirit of the religion of technology in the United States at the end of the nineteenth century. In his first book, *The Religion of Solidarity* (1874), Bellamy describes the "tendency of the human soul to a more perfect realization of its solidarity with the universe, by the development of instincts partly or wholly latent."[5] And he further explains: "In the soul is a depth of divine despair over the insufficiency of its existence . . . and a passionate dream of immortality." The "half-conscious God" that man is is called to fully recognize his divine elements. In 1888 Bellamy published a utopian novel entitled *Looking Backward.* Retrospectively, from the perspective of the year 2000, our perspective today, he depicts the United States of the year 2000 as a technological utopia, as an "ideal" society. The technological achievements that he describes have actually become reality, but whether they have generated the ideal society is debatable. Extremely enthusiastic about the emergence of technology, reflecting the exuberant optimism of the times, Bellamy even writes: "Humanity is proving the divinity within it," and characterizes these prospects, in a highly poetic manner, as "a vista of progress whose end, for very excess of light, still dazzles us."

Lem has also attempted to look back from the future in his *Star Diaries.* There he depicts the past of the future from the approximate perspective of the year 3000 A.D.:

> At first, hardly anyone and then no one was born from a union between a man and a woman anymore, but from a cell which was placed in a uterator, an artificial womb, and one could scarcely refuse the sacraments to all of humanity with the justification that they had been created through virgin birth. On top of that, the onset of the newest technology, one of the consciousness, could not be overlooked. The problem of the spirit in the machine, born through electronics and its reasoning computers, was resolvable, but then the next problem arose, that of the consciousness and the psyche in liquids. Intelligent and thinking biochemical solutions were synthesized that could be bottled and combined, and each time a new personality was created, more spiritual and intelligent than all Dichtonians (the name of the planet which the guest is visiting in the year 3000) together.[6]

Lem continues:

> There were dramatic confrontations among the churches at the synod in the year 2479 on the question whether a machine or a

> biochemical solution could have something similar to a soul until a new dogma was declared, that of the indirect creation. This purported that God had given the reasoning being which he created the power to create the intellects of the next litter. But that wasn't the end of the transformation, for it soon became evident that artificial intelligences were capable of reproduction or of synthesizing humanlike beings or even normal human beings out of a pile of matter according to their own ideas. Later attempts were made to save the dogma of immortality but they collapsed under the onslaught of further discoveries that overtook the twenty-sixth century like avalanches. No sooner had the dogma been reerected with a modified interpretation than the synthetic technology of consciousness appeared.[7]

The only thing remaining was the dogma of indirect creation.

This dogma of indirect creation is anticipated in the visions of some natural scientists today. Richard Seed considers his plan to clone human beings a realization of the true image of God in the human being, because the ability to develop God's image, encompassed within the image of God inherent in the human being, definitely must be passed down to the next generation in an improved form. In a published advertisement advocating cloning technology, twenty American scientists and Nobel laureates argue similarly, as does a group in California involved in the further development of artificial intelligence who believes that the human being is not yet intelligent enough to solve the problems that he or she produces through technological solutions. The creation of the human being with artificial intelligence should take place together with that of the technological solutions, an intellect capable of resolving the problems arising from these.

This conception of the creative power of the human being is clearly one-sided. What we see here is the alliance of our society with the future, an alliance grasped and internalized by the human spirit and human consciousness, understood as progress. I do not mean this ironically. Actually, if we consider our societies plurally, in modernity we have in a certain sense already entered an irretrievable, irreversible alliance with science, by allowing scientific inquisitiveness in every possible direction and by considering it a value in itself: an alliance with technology, that is, with the idea that that which we recognize will be converted into competence and skills, and an alliance with the economy, that is, that that which we have realized in technology will become an article of production, of commerce, of sales, and of consumption. This alliance, as I have stated, is not revocable, and it has achieved a number of fantastic and fascinating successes in our world. The fact that our world has become more comprehensible, predictable, closer, and interchangeable in culture is undeniably a gigantic opportunity that has arisen from this alliance. However, it also involves biases and dangers. They—the second characteristic of this conception of human creativity—find expression in the "breakthrough" theory (*Durchbrecherthese*). This theory states that the human being will always be capable of solving the problems that he or she has created in solution of other problems. There are well-known examples that would question the accuracy of this contention, for example, in the field of atomic energy. Can we solve the problems that we have created through this particular solution to the energy problem, the question of atomic waste disposal and other related issues?

This kind of development of the creative power in the human being, which I have characterized as one-sided, involves a very specific understanding of freedom. This is freedom as self-determination or choice—informed consent and free choice—without a predetermined hierarchy of goods. It is a choice without metaphysics, without ontology, and without anthropology, a choice based on advantages and disadvantages. These advantages and disadvantages, however, are in turn self-determined. And this is a creative ethics of freedom corresponding to the conception of breakthrough futurology. Crucial is what kind of creative ethics of freedom we recognize. For, as Bernd Gräfrath establishes in his study *Es fällt nicht leicht, ein Gott zu sein: Ethik für Weltenschöpfer von Leibniz bis Lem,* when the human being has himself or herself become a creator of worlds, and no longer stands passively in awe before the world creation of a God-creator, then the question that preoccupied Leibniz will become a question directed toward the *human being*. How can the world in which we live be seen as the best of all possible worlds?[8] This question directs itself not only toward God with regard to the first creation process but also toward the *second* creation process, during which we really pervade our world. Karl Rahner called this *hominisieren.* Will not the metaphysics of creation, in which God was the primary agent, be replaced by a morality in the sense of a creative ethics of freedom, in which the human being is the primary agent? I would like to clarify this problem with the assistance of a few quotations. For example, from Albert Einstein: "When I evaluate a theory, then I ask myself whether I would have set up the world in this way if I were a god."[8] Bernd Gräfrath formulates it this way: "Leibniz's theodicy cannot only be read as a metaphysical explanation of the world but also as an ethics of creation. . . ."[10] And Ulrich Steinvorth comments:

> Like Plato's demiurge and Leibniz's God–creator we, too, must decide today what the best of all possible worlds is. Which possibilities are worthy of being realized, yes, if any possibility should be realized at all or if something should be rather than not be. Modern technologies force the age-old metaphysical questions on us again as moral questions.[11]

This is visible in the contemporary development of scholarship and research. Today philosophy is rapidly transforming itself into the philosophy of science and ethics, and theology is transforming itself into ethics and spirituality. There is also an ambiguity in the present boom in ethics, since the boom in ethics involves a sort of world management. Can we accept a moral world management, a creative freedom of this kind, without learning to distinguish between what truly serves the human being and what can truly deliver him or her? Morality as the securing of the power of deliverance of modern technology is found everywhere, but a morality that comprehends the human being as a finite being is becoming increasingly rare.

The Forgotten Finitude of the Human Being

In the conception of the creative freedom of the human being as free choice, in the self-determined life, there is hardly a hint of the fact that the human being is a being that can die. Or that the human being is an individual being, dependent on others, or that the human being always has only partial power over himself or herself, never total power. Nor is there recognition that the human being is a being capable of making mistakes, and that this fallibility also means that we cannot simply reverse or negate the consequences of our actions.

When we create something, it develops a momentum of its own, and we cannot control the consequences of everything in which we have invested. This is the case with the children we have produced. The consequences of our actions toward them are not simply reversible if, when they are forty years of age, we attempt to reconvert them to that which we originally had in mind for them. This is impossible. Certainly, we can develop cultures within our natural predilections, but we are not the lords of evolution. We are subject to the laws of evolution even when we shape evolution. In the case of so-called auto-evolution, we are only capable of reproduction but not, for example, of a production, of a beginning *ex nihilo*, out of nothingness. For example, if we say that we have created human life from a test tube, this is not correct, because we have not created the conditions requisite for this to succeed. We have only reproduced them; "reproduction" is the corresponding word that exposes our finitude.

What I have said here is perfectly obvious. Philosophy is essentially nothing other than the exposure of what is obvious, evident. It is obvious, it is comprehensible, to every human being that we are finite beings. How can we forget this? We can forget this. In contemporary philosophy we find references to the concept of finitude, for example, in the works of the postmodern philosopher Richard Rorty, but he uses the term in the sense of chance. He means that something does not proceed in accordance with the laws of causality but that chance occasionally plays a trick on us, and he analyzes this philosophically. An anthropology of finitude is not to be found in contemporary philosophical discourse. There are occasional exceptions such as Hermann Lübbe's theory of religion as a *Kontingenzbewältigungspraxis*, a means of coming to grips with contingency. Perhaps theology can inspire philosophers to contemplate this question, if philosophers were still willing to read theological texts today.

If they did, they could discover what they are looking for in an essay by Eugen Drewermann, for example. In his article on finitude ("Endlichkeit") in the *Neues Handbuch theologischer Grundbegriffe,*[12] Drewermann characterizes belief as a twofold motion (*Doppelbewegung*). The human being who has become self-conscious frees himself or herself from finitude in an endless resignation experienced through finitude, and then regains this from God on the basis of the infinite. "It," he writes, "necessitates considerable faith to overcome the fear of freedom that repeatedly compels one to search for spurious reassurance in the one-sidedness of existence and to escape to a pole of the tension of being between finiteness and infinity instead of seeking a true synthesis between finiteness and infinity."[13] Indeed, we attempt to live the infinite dreams that we have in a finite existence in such a manner that we forget that we dream them as finite beings. This tension, if we forget it, plunges us back into finitude because we experience our mortality all the more, all the more drastically, our fallibility, our guilt, our helplessness in the face of events like those that took place in Kosovo in 1999. "This is because," Drewermann concludes, "there is no longer a God who would fill the existential gap of the nonnecessity of the finitude of the human being through his positive will of existence."[14] The existential gap, the gap of being (*Seinslücke*), is one way of putting it. What has disappeared is the question about the condition determining the possibility of our existence as we exist. I do not wish, nor presumably does Drewermann, to place God in a gap. This would be a typical theological procedure: one discovers a place where something is missing, places "God" there, and purports that this proves his existence. I think that this gap is only painful from the perspective of belief—Drewermann says this, too. Belief lives in the overflowing self-communication of God, and this is why Drewermann sees the gap here, the gap of finitude.

Why is this insight so important for freedom? I have spoken of a creative freedom that understands itself above all as *choice*. As self-determined beings we have the choice between various alternatives. Yet if we conceive of ourselves in finiteness, in dependency, in mortality, in fallibility, then we know that we are located in a movement based on an experience, the experience of limitedness. My experience of limitedness is that I have a body, that I am a body, which is finite, which has a certain location where it is immovable, a specific flesh, a specific build, and that I have only limited time, not only the limited here of space but the limitedness of time, of death. Only a freedom that arises from this conception of finitude has the character of liberation. For experienced finitude in this sense is a recollection of creation, thankfulness for the contours of existence that my body gives me, thankfulness for the intensity of the limited time that allows me to experience time in its intensity at all, that makes the moment become so important.

There can be gratitude for a form of responsibility that does not simply consist of choice but constitutes an answer to something given. I would express this insight as *felix finitudo*, blissful finitude. This is a new world juxtaposed against the one-sided world creation that I sketched at the beginning; it arises out of the praise of finite freedom as the essential creative power in the human being.

Heinrich Rombach has summarized finite freedom as the basis of creatorship in one concept: *Konkreativität*, literally "concreativity" or "co-creatorship."[15] Meister Eckhart has mentioned, as has Martin Luther, that "one should learn to act together with his God."[16] *Cooperatio* can be polemically separated from *concreatio*, as Luther did. In both *The Religion of Technology* and *Ethik für Weltenschöpfer,* much of the discussion focuses on the autogenesis and auto-evolution of the human being. Stanislas Lem speaks not without ridicule of the "polyverse of omnigenerative creationism."[17] This is a poetic phrase. Lem is obviously playing with the poetic power of technological language. When everything is creative, nothing is creative any longer but creationist. Persons who could be affected by anything retrospectively are no longer responsible, and only that is decisive. Following the unstoppable sequence of the "indirect creation" of a human being, who uses his likeness to God autocreatively, according to Lem, there is no longer a universe but a "polyverse." Everything is generative; everything is creative; but nothing is a person anymore. If we continue to hold that a personal concentration is inherent in the experience of the creative, then we must insist on the co-creatorship of the human being and his or her co-action. What does this mean? If I comprehend myself as a fellow creature or co-creature, ensuing from the creation (God's first creation), and at the same time as a co-actor, co-agent, in the realization of this first creation of God, then I must discover a link to God's acting in the world insofar as it is revealed to me through my own finitude.

This kind of creation relationship also opens another perspective on the responsibility that emerges today given the fact that the human being has placed so much hope in deliverance through individual development. It is not a matter of replacing metaphysics with morality but of correctly determining *the ethical* on the basis of the co-action. Morality does not give our action its ultimate meaning; instead, we know that our action cannot be merciless, that it cannot be blinded in self-importance, and receive a new incentive for this type of praxis, in our awareness that not only the actions of the human being affect him or her retrospectively but the motivation behind the action as well. In this manner, a conception of creatorship emerges that is totally different from that of self-determination through choice, the slogan of the American Pro-Choice movement.

A Literary Reflection on Finitude

In the fifth act of *Faust,* Goethe has portrayed how Faust appears as a demiurge for the well-being of the human race (he attempts to regain land for new settlements from the sea). Faust sees how Mephistopheles and his cohorts begin the task of canalizing new land. A tableau is presented at this edge of the sea: a tiny house stands on a dune, surrounded by linden trees. Philemon and Baucis, a frail, elderly married couple, familiar to us from Greek mythology, live in this humble dwelling. A Christian chapel stands next to the house. The chapel bell rings, and the two elderly persons go to prayer hourly. This monklike hermit's life as a pair is a lovely vision (for couples, too), but what is relevant in this context is this: it is an idyll, an idyll to which, as described at the beginning of the scene, a wayfarer is admitted. He recalls being washed ashore there and saved by this same couple long ago when the sea was still untamed. The Good Samaritan appears characteristic of this world. It is unclear whether the wayfarer has something of a godly messenger about him.

Faust's problem is that he needs the couple's hutlike dwelling for the technological revolution he plans for the benefit of the human race, and he offers to build them a better domicile elsewhere. But, as the case often is with older persons, they prefer to pray and trust in their "old" God (this is Goethe's choice of words), and Faust becomes annoyed. Faust speaks of justice and his goodness while his heart "rages," because it is attuned to progress. Without thinking, he loses patience and proposes that Mephistopheles use force to achieve that which he himself cannot carry out with the consent of the elderly couple, namely, to relocate them in a new and beautiful place. The subsequent scene depicts how the tower lookout, "Zum Sehen geboren, Zum Schauen bestellt" ("born to see, appointed to watch"),[17] sees Mephistopheles and his wild band of men set fire to the couple's hut. They both die in the fire; only the wayfarer offers resistance and dies fencing. Mephistopheles offers his own version of the events, alleging that the elderly pair suffered heart attacks after he broke down the door to get in.

Faust is confronted with this situation, and he actually experiences conflicting feelings. A colossal vision of human progress, he is forced to admit, requires sacrifice. His solutions to problems create new human problems of this kind. In this inner conflict "Care" comes to light. Four gray women, "Want" (*Mangel*), "Debt" (*Schuld*), "Need" (*Not*), and "Care" (*Sorge*) appear. The crones Want, Debt, and Need do not gain entry to Faust's palace but Care sneaks in through the keyhole; in other words, she is already there. She is already present in the twinges of self-conflict deep within his heart. Care informs Faust of human finitude. For Care is only another name for the finitude meant here.

Faust is informed of finiteness, and he recognizes its existence but does not accept it. He recognizes that this kind of world creation, as he has planned it, cannot be realized since he is full of "care," his own finiteness. But he refuses to accept this, declaring like a creed: "Ich werde sie [deine Macht] nicht anerkennen" ("I will not accept it [your power])."[19] Consequently, the crone Care blinds him. Faust is now blind, and in this state of blindness his true condition is revealed. This is the final scene in this sequence: Faust does not perceive that the engineers and workers are not digging canals but are in actuality digging his grave. Although he hears the digging, he is not aware of its true meaning, a situation which Paul Celan also alludes to in his poem on (grave-)digging.[20] While Faust simultaneously proclaims his inner vision of the progress of the human race, and thus gladly says to the moment "verweile doch, du bist so schön" ("tarry awhile, you are so lovely," verse 11582),[21] he is greatly mistaken, because this is the moment when his grave is being dug and not the moment when a brave, new world emerges.

Mephistopheles, facing the audience, makes fun of the blind man, ridiculing the way he has failed to realize the insights of his own human reality. Faust dies without having learned to see; he dies as a blind man, recalling, in the language as well, the passage on blindness in the Gospel of John where it is precisely the seeing, or those who purport to be the seeing, who are blind, and where one can paradoxically say that true sight and insight are only possible through blindness.[22]

At the end of Faust's burial, the scene that essentially concludes the drama, all that remains is the insight into human finitude, nothing more. What follows—the delightful game between devils and angels, during the course of which Faust's soul is taken to heaven by the Mothers—is perhaps unimportant. In the age of Hollywood a "musical" at the end can only be described with a certain reticent irony. It is crucial that this tragedy—it is undeniably a tragedy—at the end of *Faust II* is the key to insight into finite freedom. For only the tragedy—this is Goethe's message—can generate a change of heart or a new consciousness. It is not consolation, even though we need it, that transforms us. Rather it is the confrontation that devastates us, that makes something of the *tremendum* and *fascinosum* visible, and that thus makes clear to us to what degree we are "concreative," that is, to what degree we can grasp this *fascinosum* of a recollection of creation and reinforce it through our intellect. Only in this respect are we open to a new perspective on that which we create as human beings. We are not only technologists and poets but also practitioners who are aware of what happens to ourselves when we act.

Chance or Self-Determination?

What Goethe's *Faust* shows us is succinctly presented by Hannah Arendt in her famous book *Vita Activa*.[23] Modernity effects a reversal of medieval thought in its comprehension of "nature" no longer as given but as relinquished. The future becomes the space of human design; the present, its arena of experimentation. The past is considered exemplary insofar as it envisions or even anticipates later steps. Francis Bacon early on rejected the idea of the old as authoritative, viewing it rather as the way to the new. The conception that the "new" will become just as outdated as the old, and that there is essentially nothing "new under the sun," the subject of contemplation from Ecclesiastes to the waning Middle Ages, was forgotten.

Had the fathers of philosophy and theology still considered the intervention in nature and the pure action *sub specie aeternitatis* as unfruitful, had they comprehended praxis above all as *ethical* praxis retrospectively affecting human beings, praxis now came under the spell of success. Success replaces the teleological order of nature. However ambiguous success may be, however unachievable for all, however prone to failure, modern belief in progress remains linear, discerning an ascending line from earlier times to the present, from the present to the future. The *vita activa* is freed from dreams of virtue, compassion, and the transmission of truth; now it is the quintessence of feasibility. The speed of progress increases, and the "contemplative" or reflective powers cannot keep pace. Here the ethical maxim is also valid teleologically: one should not solve problems in such a way that the problems arising from the solutions are greater than the problems being solved. The use of nuclear energy and the potential of biological and chemical weapons to date show how extensively this analytically evident criterion has been ignored when conquering new fields. The risk-loving society has become accustomed to acting despite uncertainty, in expectation of being able to deal with problems if they should actually arise.

Given progress that believes strongly in light in the dark and that believes all the more secularly as metaphysical forms of belief recede, those values that a humanistic culture struggles to develop and maintain have little chance of survival. Several examples:

The cultural assessment of natural chance is ambivalent. On the one hand, this is a source of chaos and destruction. Natural catastrophes confirm this again and again. On the other hand, natural coincidence is also a source of passive happiness. For this reason we have left the natural state of the human being outside the realm of our intervention and planning up to now. With early genetic diagnosis (PND, PID) we have begun to change this. The end of "natural" culture seems connected with the end of the worldwide acceptance of natural law that is not totally preserved in the individual human rights. As Jürgen Habermas has observed, in this context the "ethics of genus" (*Gattungsethik)* that gave every human being the same rights "on the basis of nature" disappears. What is contested, in the eyes of Cardinal Karl Lehmann, is "the right to be a human being,"[24] whether the human living thing is transformed into a medication or rejected on the basis of genetic information.

The technical possibility precedes the legal permissibility in such cases. There the systemic alliance in biotechnology works complementarily: genetic engineering, reproductive medicine, and clone technology produce both the new options of predictive and regenerative medicine together. Common to this procedure is the in vitro situation: the human embryo produced from the extracorporeal fertilization, the status of which immediately became the object of dispute (on the one hand, scientific data requires interpretation; on the other hand, no other turning point comparable to the fusion of egg and sperm cell is to be found). The "in vivo genesis" of the human being is thus not only subject to intervention and planning, it also exposes the unprotected embryo outside the womb. This embryo already contains the initial physicality and the particular sex. Its extradition to the petri dish corresponds to the betrayal of the greatest intimacy of the human being. The warmth and security for the child and the warmth of the bed are rigorously separated. The physical symbolism of the unity of the sexual womb with the maternal womb is abandoned. Some will say that this was already the case with the regulation of conception. And some will consider the doctrine of the Catholic Church prophetic. But between the intentional unity of relationship and fertility in every act of sexual intercourse and the total separation of sexual relationship and procreation there is again a crucial difference. The central idea that I wish to present here is the loss of physicality in the relationship, as also evident in the reduction of the human body through information technology simultaneous with its utilization as a keyboard of attractions.

Whoever does not respect the body has long lost respect for the belief in resurrection. Immortality, however, is, like metempsychosis, a positive metaphor. Immortalized through the artificial intelligence of my individual type, copied by the computer, or immortalized through the exchange of my cells ad infinitum; whatever kind of world this world will be, it will not be a just world nor a world for all since I nourish myself from the instrumentalization of others. The metaphor of "cannibalism" may be exaggerated, yet the other is no longer a coguarantor of my dignity but merely an instrument that I scrutinize and utilize depending on my greed for success. Evil roves, in the metaphorical language of the psalmist, "like a roaring lion, looking for someone to devour."

Particularly affected by this disregard for the integrity of the body are women. Assisted reproduction is carried out on the woman's body alone. She bears the risks for the infertility of the man. She contributes through surgical intervention the egg cells that become resources and lead living human things to the stockpile for destructive experiments. She endures prenatal diagnostics bodily. She reaps the problems sown by others in their solutions to problems. She serves as the most extreme example of need and desire, presented by researchers and medical doctors, who see themselves in the role of helpers and healers, that is, a savior role, yet at the same time fear the invalidation of their knowledge if they are unable to propel it toward new insights through experimentative utilization. The freedom of research, arduously achieved as the freedom of the spirit of inquiry, becomes the freedom from fear of invalidation because without the exploitation of its insights in the name of utility, science could not develop at such a rapid pace. Accelerated action takes the place of contemplative prudence.

All of this is justified by self-determination. Self-determination is the instrumental word for the breakthrough of science, technology, and economy. However, this word and its actual function contradict one another. The functional meaning of self-determination is atomistic individualism that substitutes information for counseling, and bureaucratic consent forms for relationships. The paradox of self-determination lies in its regulation from the outside. There the decision is made whether the woman belongs to the percentage of those justified in exercising self-determination. And there the content of the offer of self-determination is decided upon as well: which offer do we make to whom?

Theoretically, there could be a different offer—an offer for others. There could be alternatives. But no one thinks seriously about alternatives in offers or in the needs of persons. The "dictatorship of the genes"[26] swallows large segments of potential medical progress in other fields. Will other alternatives in Parkinson therapy continue to be explored and financed if many tend to the stem cell option? Countless other examples come to mind.

Self-determination is not to be challenged as a criterion. But the following questions must be raised again and again: which self-determination, within which limits, to whose advantage? Self-determined self-determination is self-committing self-determination, true auto-nomy. The self becomes exceedingly aware of its own remoteness from others and the unavailability of the other. The word "dignity" does not now constitute a summation of values but establishes inaccessability (following Immanuel Kant). Here it is important to differentiate self-determination. "Know thyself" was the motto of Greek philosophy. A cultural amnesia has transformed this into "make the best of it." But what should the best be made from? Is optimization not blind to the "it" that it optimizes, and even blind to the persons whom it involves?

Self-determination is often coupled with the conception of not having to bear "unnecessary" suffering (a formulation of the Council of Europe in 1977). But does the current level of technology determine what a need is, what solidarity is, and so demand the involvement of the other?

Dealing with suffering has become a decisive question of the right culture. The spiritual alleviation of suffering through the cross, a double-edged undertaking that can connote assistance as well as oppression, has long ago been discovered. Suffering cannot be sought—it is to be survived. But here survival does not always mean action. The enemies suffering and death are not to be defeated by force. They are defeated when they lose their power in the face of the fundamental passivity of the human being of faith. This is no excuse for inactivity or for the failure to lend assistance (nonfeasance) in the face of suffering and distress. Incidentally, this thought leads us in the terminology of rights and the distinction between "negative" rights and "positive" rights. We have the right that no harm or suffering is forced upon us; we do not have the right to make others the instrument for the fulfillment of needs, the nonfulfillment of which causes us to suffer. Nor are the others instruments for the reduction of suffering.

We have, in Alan Gewirth's opinion, the right to the necessary goods, and these must be firmly established as a priority in a "community of rights."[27] We do not, however, have the same right to "additive goods," through which others are instrumentalized to the point that their existence is threatened, nor to goods that call for a solidarity in which the goals are comprehensible but the means no longer justifiable.

The Loss of Acceptance as "Structural Sin"

Today we suffer from our incapacity to accept other beings without reservation. The rejection of others once was comprehended in terms of *personal* guilt but has today become *structural* sin: the loss of unconditional acceptance. This is just as visible in our relationships as in the expression "(Un-)Zumutbarkeit," translatable as un-reasonableness or lack of deliberation and premeditation, which has become a central term for immunity from criminal prosecution. (This applies, for example, in the case of abortion, in some legal systems.) Of course, to force responsibility via prosecution would be wrong, from the moral viewpoint as well. To utilize individual lack of reasonable judgment and responsibility in order to avoid societal problems is cynical. The structural sin lies in this cynical variant of reason which replaces solidarity and assistance with the assignment of guilt to individuals. Their psychic suffering elevates the economic society with its deficiencies in solidarity to the sovereign over the norms of constitutional society, which actually should make human life acceptable, inviolable, and reasonable.

The loss of unconditional acceptance is at the same time a cultural amnesia, a structural sin, and a distortion of God. For God has allowed his history with humankind to be written as a history of acceptance, a progressive purification of the image of God, until God and the human being were bonded together in this faith of acceptance (here we could compare the Pauline lines on love). God is not thinkable without this longing, taken from, yet exceeding, human powers; the human being is not thinkable without this human image of acceptance and without its base in God. Together with the amnesia of acceptance, advocacy for the benefit of those who cannot themselves decide and are vulnerable has lost its cultural *Sitz im Leben*. The human has been segmented, which determines the nature of the advocacy. Not the advocate but those to be represented by him or her are "appointed" as human beings, prehuman beings, or "no longer human beings." Having lost its structural roots in the unity of the human being and human dignity, forced to revert to the competition of attributes, advocacy loses momentum. (Not even affected handicapped or disabled persons are necessarily represented on a national ethics commission.)

Cultural amnesia exists on the basis of a false choice between remembering and forgetting. Our technological advances are drummed into our heads day in and day out, by the media and by the constant training of our capacity for knowledge; consciousness of our failure with respect to laboriously achieved values is only occasionally present. Ignoring memory produces the normative power of the fictive, as if the promised were the true reality. Whoever promises a cure is right from the very beginning. A capitalist teleology replaces the communist. The present becomes bearable through the prognosis and the fiction of its future. And when fictions are redeemed, the world immediately divides itself into those who participate and those who are excluded.

Fiction as a worldly teleology is encountered only with skepticism by followers of a different faith. The unbelieving belief watches while the believing unbelief expands. Idolatry is based on a new affirmative culture of the fictive. This culture has long become a part of the structures of knowledge, learning, working, and pleasure. Fantasy and cyberspace allow the memory of history only as a citation of robes and weaponry. If history is also based on narratives, at least these are recollections and not exclusively constructions that draft the present on the basis of a fiction of the future. In conjuring the imperfect tense, the imperfect, too, has a chance.

We have identified the common structural trait of the cultural assessment of losses as forgotten finitude. It is probable that we will experience finitude as a dictate of the real future. Chernobyl was one such experience, September 11, 2001 is another. If the "perfect world" once again becomes a world of the imperfect, we will learn to better deal with the possibilities and limitations of human freedom.

Translated by Jo Ann Van Vliet

Notes

1. In the following essay, I will not discuss the literary dialectics of the treatment of myth, which Thomas Mann called adaptation or transformation (*Umfunktionieren*) and which can be characterized as a demythologization of myth as well as a mythologization of the demythologized. See Dietmar Mieth, *Epik und Ethik: Eine theologisch-ethische Interpretation der Josephromane Thomas Manns* (Tübingen: Niemeyer, 1976), 36–49, 148–70. For my contribution to bioethics, see Dietmar Mieth, *Was wollen*

wir können? Ethik im Zeitalter der Biotechnik (Freiburg im Br.: Herder, 2002).

2. *Goethe Werke, Jubiläumsausgabe,* ed. Albrecht Schöne and Waltraud Wiethölter (Darmstadt: Wissenschaftliche Buchgesellschaft, 1998), 3:51, *Faust I* (verse 1335).

3. *Designing Life? Genetics, Procreation, and Ethics,* ed. Maureen Junker-Kenny (Brookfield, Vt.: Ashgate, 1999). David McDonnell authored the essay entitled "The Machine in Man," in *Designing Life?* 64–69.

4. See esp. in *Sterntagebücher* (Frankfurt a.M.: Suhrkamp, 1978), 212–74.

5. This and the following quotations from Bellamy's *The Religion of Solidarity* are cited from Noble, *The Religion of Technology,* 98–99.

6. Since the English translation of Lem's *The Star Diaries* was not available on the European continent, these passages have been translated from the German edition, the *Sterntagebücher.*

7. Lem, *Sterntagebücher*, 226.

8. Gräfrath, *Es fällt nicht leicht, ein Gott zu sein,* 17–71.

9. As cited by Albrecht Fölsing, *Albert Einstein: Eine Biographie* (Frankfurt: Suhrkamp, 1995), 791. See Gräfrath, *Es fällt nicht leicht,* 67.

10. Gräfrath, *Es fällt nicht leicht,* 17–71.

11. Ulrich Steinvorth, *Warum überhaupt etwas ist: Kleine demiurgische Metaphysik* (Reinbek: Rowohlt, 1994), 115; cf. Gräfrath, *Es fällt nicht leicht,* 67–68.

12. Peter Eicher, ed. (Munich: Kösel, 1991) 1:332–96.

13. Ibid., 336.

14. Ibid., 337.

15. See Heinrich Rombach, *Strukturanthropologie: Der menschliche Mensch*, 2d ed. (Freiburg/Munich: Alber, 1993). See also Karl Rahner "Experiment Mensch. Theologisches über die Selbstmanipulation des Menschen," in *Die Frage nach dem Menschen: Aufriß einer philosophischen Anthropologie. Festschrift für Max Müller zum 60. Geburtstag*, ed. Heinrich Rombach (Freiburg/Munich: Alber, 1966), 45–69. Peter Sloterdijk uses this quotation from Rahner without taking Rahner's moral reservations into account ("Der operable Mensch," in *Der (im)perfekte Mensch: Vom Recht auf Unvollkommenheit*, ed. Stiftung Deutsches Hygiene Museum [Ostfildern-Ruit: Hatje Cantz, 2001], 97–116).

16. Meister Eckhart, *Die Deutschen Werke*, ed. Josef Quint (Stuttgart: Kohlhammer, 1936–), 1:291.

17. Lem, *Sterntagebücher,* 237.

18. Goethe, *Faust II* (see note 2), 379 (verses 11288–89).

19. Goethe, *Faust II*, 385 (verses 11493–94).

20. Paul Celan, "Es war Erde in ihnen," *Die Niemandsrose* (1963) = *Gedichte* 1 (Frankfurt: Suhrkamp, 1978), 311. On Celan's poem, see

Regina Ammicht-Quinn, *Von Lissabon bis Auschwitz: Zum Paradigmenwechsel in der Theodizeefrage* (Freiburg, Switzerland: Universitätsverlag; Freiburg, Germany/Vienna: Herder, 1992), 9–11.

21. Goethe, *Faust II*, 389.

22. This motif is also found in José Saramago's novel *Blindness* (London: Harvill Press, 1997). See Regina Ammicht-Quinn, "Versuch über die Blindheit," *Interdisziplinäre Ethik,* ed. Adrian Holderegger and Jean-Pierre Wils (Freiburg, Switzerland: Universitätsverlag; Freiburg, Germany: Herder, 2001), 269–87.

23. *Vita Activa oder Vom tätigen Leben*, 6th ed. (Munich: Piper, 1989) = *The Human Condition* (Chicago: University of Chicago Press, 1958).

24. See *Die Zukunft der menschlichen Natur: Auf dem Weg zu einer liberalen Eugenik?* (Frankfurt a.M.: Suhrkamp, 2001).

25. Sermon to the German bishops, 2000.

26. See Dietmar Mieth, *Die Diktatur der Gene: Biotechnik zwischen Machbarkeit und Menschenwürde* (Freiburg i. Br.: Herder, 2001).

27. Alan Gewirth, *The Community of Rights* (Chicago: University of Chicago Press, 1996).

PART I

PROJECT CONTRIBUTIONS

Section II: Moral Experience, Evaluation, and Virtue

4

The Need for a Dynamic and Integrative Vision of the Human for the Ethics of Genetics

KEVIN T. FITZGERALD

Talk about the future promise and peril of genetic technology is constantly before the public. Newspapers and television continually have segments announcing recent advances in genetic research. Even scientific journals have regular reports on aspects of the issue that are not specifically scientific.[1]

Genetic technology offers the amazing opportunity to modify human genes in order to combat genetic disease. However, if it is possible to alter the human genome in an attempt to improve health, it will also be possible to alter the human genome in an attempt to effect a fundamental change in human nature. Apart from the scientific issues of what will and will not work, such potential for change raises ethical questions regarding which genetic interventions should be pursued and which should be left unattempted. These questions need to be answered with regard to the impact genetic interventions will have on the individuals directly treated, on their progeny, and on society at large.

Can genetic interventions alter the nature of human beings?[2] In order to answer this question, an understanding of humanness and what, if anything, is crucial to being human will be central to the discussion. Applying this understanding to the ethical questions raised by human genetic interventions will provide the basis for deciding what should or should not be attempted in altering human beings.

Concepts of humanness, and the related concepts of what is good for human beings, are, at least implicitly, situated within more comprehensive theories of human nature or "philosophical anthropologies." Being comprehensive in scope, these anthropologies always rely, in part, on the scientific knowledge of a given time. Problems can arise when a given philosophical anthropology is based on dated or inaccurate biological information.

Problems arise because the contemporary scientific knowledge that makes genetic interventions possible may challenge or contradict the ideas of what is normal or good for human beings found within a given philosophical anthropology. Hence, if a scientifically flawed anthropology is used as a basis for the ethical analysis of human genetic interventions, the conclusions concerning which genetic interventions to try and which to avoid may be at odds with the best scientific knowledge of the time.

This essay will compare some of the different types of philosophical anthropologies often employed in the ethical analysis of human genetic interventions in order to elucidate the type of philosophical anthropology best suited for this ethical analysis. The philosophical anthropology of Karl Rahner, a German theologian, will then be suggested as representative of this preferred type. However, for this comparison of the different types of philosophical anthropologies to be clear, a more detailed account of the current state of ethical analyses of human genetic interventions is required.

A commonly used device in the ethical analysis of human genetic interventions has been to employ a grid using a therapy/enhancement distinction on one side and a somatic/germ–line distinction on the other. This grid then allows one to divide human genetic interventions into four different types for ethical analysis [see figure 1].

Figure 1

	Therapy	Enhancement
Somatic	1	2
Germ line	3	4

The value of the grid rests on the applicability of its distinctions and the relevance of its four types of human genetic interventions to the ethical analysis. Up to the present time, a broad consensus of national and institutional panels have supported the careful pursuit of somatic cell gene therapy (type 1) while concluding that pursuit of the other three types of genetic interventions—involving either nontherapeutic genetic changes or changes that can be passed on to future generations—is not yet advisable. This broad consensus in support of only type 1 interventions, however, may not last much longer because of advances in genetic research that are undermining the relevance and applicability of these distinctions.

The blurring of these distinctions is particularly clear with regard to differences between therapy and enhancement. One example of how recent research blurs the lines between gene therapy and genetic enhancement can be seen in the variation in susceptibility to HIV infection found in different people. Reports of various individuals' resistance to multiple exposures of HIV from around the world combined with some basic research has led to the conclusion that certain people have genetic differences that make it much more difficult for HIV to infect their T-cells (the cells of the immune system the virus attacks).[3] Using this knowledge, one could argue that large-scale genetic interventions should be pursued to make people more resistant to HIV infection, especially for those who are at high risk of exposure such as surgeons working in areas of high disease incidence.

As mentioned above, internationally accepted guidelines only allow for disease-oriented gene-transfer protocols to be pursued. Since vaccinating people to produce disease resistance is already a mainstay of modern medicine and since HIV infection is so devastating, one could argue that a genetic intervention to prevent HIV infection should be considered a gene therapy and not a genetic enhancement. In rebuttal, another might argue that the people who would be receiving the treatment are not ill and do not require the treatment in order to avoid the disease. In addition, different people having different levels of genetic resistance to different diseases within a given population is the natural condition. Hence, genetic interventions to prevent HIV infection must be considered an enhancement. This potential debate illuminates the problems that may soon be encountered when using the therapy/enhancement distinction with respect to human genetic interventions.

In general, whether or not a genetic intervention will be considered therapeutic or enhancing in character will depend on how members of society and the medical profession interpret the goals of the intervention and the condition of the patient. For example, the same genetic intervention for increasing intelligence could be viewed as therapeutic when applied to an individual evaluated as mentally deficient but considered an enhancement if done to an individual of "normal" intelligence. Hence, answers to the question of whether or not a genetic intervention is to be considered a therapy or an enhancement lose their relevance if members of society and the medical profession do not agree on answers to the question of what is to be considered health and what disease. Since health is usually defined by reference to the "normal" functioning of the organism, the key question, then, for societal or medical determinations of what is health and what is disease is: How will "normal" human physiology and behavior be delineated?[4] Any adequate answer to this question will require a philosophical anthropology that is responsive to the challenges new genetic information is constantly raising regarding current concepts of normal human functioning.[5]

From the above examples, one can readily discern the rapidly increasing problems with the therapy/enhancement distinction. These problems stem from the reality that as new genetic information and technology challenges traditional ideas as to what is healthy, normal, or good for human beings, then making distinctions, such as therapy versus enhancement, which are based on these ideas will become more arbitrary and less useful. In a similar manner, the increasing number and complexity of the different options for human genetic interventions will create a problem for ethical arguments which rely on the somatic/germ-line distinction.

There are fundamentally two aspects of germ-line intervention that are cited in support of using germ-line genetic interventions.[6] They are efficiency and effectiveness. The value of these aspects is not unique to germ-line genetic interventions. They are common goals for many medical pursuits, so the question here is one of evaluating just how much benefit these two aspects of germ-line interventions supply. For treatment effectiveness, if interventions are made so that all the cells of an individual carry the alteration, then the genetic change can affect all the tissues and organs of the body. This effectiveness is particularly desirable for treating diseases that affect a variety of tissues within the body, or for treating tissues that are difficult and dangerous to access, such as those in the brain and central nervous system. As for efficiency, since the gametes of the treated individual are also altered, the genetic treatment is efficiently passed on to all the subsequent generations and does not have to be repeated for each affected individual in each generation.

To counter these obvious benefits, critics of human germ-line interventions point out several significant deficiencies in the germ-line approach. Many of the objections to germ-line interventions refer to an extremely high safety requirement, much more than for somatic cell interventions. After all, if something goes wrong it could affect every tissue in every descendant of the initial patient. However, such arguments do not determine whether or not germ-line interventions are right or wrong per se; they only set the risk standards for the procedure. Hence, the distinction here is really only one of degree, with the germ-line interventions requiring higher standards because more people can be affected by one treatment. Since the levels of control and predictability for genetic interventions are continually improving, this safety criticism may soon become less decisive for resolving the issue of where to draw lines between appropriate and inappropriate goals for human genetic interventions.

A variation of the germ-line intervention safety criticisms is the objection that humankind is not capable of employing such a powerful technology without it resulting in horrible abuses, if not tragic disasters. Somatic cell interventions are tolerable because any misuse of the technology remains confined to the treated individuals only. Since germ-line interventions could potentially affect so many people in the future, such interventions are too risky.

As far as all these potential risks are concerned, certain aspects of genetic technology research could provide sufficient fail-safe or repair mechanisms to avoid widespread tragedy and abuse for both somatic and germ-line approaches.[7] Hence, again, the real issue to be discussed regarding this objection about safety is the concept of human nature and society that lies behind the fears articulated concerning the human inability to control powerful technologies like germ-line interventions. Only an explication and evaluation of the anthropologies that undergird such objections will result in an adequate assessment of the validity and relevance of these objections.

The final objection to human germ-line interventions to be considered asserts a human right to inheriting a genome that has not been artificially manipulated.[8] This assertion places a high value on the naturalness or randomness of a human being's genome, or on the need for society to not genetically engineer future generations to meet current expectations of health, normality, or excellence. One can readily conceive of a situation where these desired values might conflict with other aspects of human existence also given great value—such as life itself.

What if one's naturally inherited genome contained a mutated gene which if left unaltered would most probably kill the human being before birth, or before the possibility of any successful intervention? Is the naturalness or randomness of one's genome more important than one's life? If not, then where will the line balancing one's health with one's genomic naturalness be drawn? In light of the potential technological advances which may make interventions to return a mutated gene to its "normal" state possible,[9] sustaining an argument for preserving the naturalness or randomness of the genome one is to inherit may be difficult. Difficulties arise for this argument because it must be made in light of what modern science reveals about humanness and what fundamental biological interventions modern technology makes possible. Whatever societies decide, the debate over this issue should involve evaluating the philosophical anthropologies used as a foundation for asserting the need always to keep inherited genomes unaltered.

The argument presented in this paper for the need to scrutinize and evaluate these underlying philosophical anthropologies is not without its own challenges. Traditionally, philosophical anthropologies have been seen as attempts to provide a complete framework for pulling together information and concepts about human nature from the various disciplines that investigate it. Using this framework, conclusions have then been derived as to which human characteristics are most significant or unique, which methodologies are appropriate for discovering these characteristics, and which actions or ways of living are best.[10] In the past century some philosophers have raised serious questions and challenges about the role and function of philosophical anthropologies in ethics.

Many trends in contemporary philosophy make the delineation and use of any "philosophical anthropology" difficult and even undesirable. In the book *Human Beings,* David Cockburn points to three particular approaches inimical to the inclusion of the notion of human beings in philosophical thought: (1) that it is almost like employing racial prejudice to include the notion of a human being in moral thought; (2) that artificial beings should be considered "persons" as readily as humans are; and (3) that one's thinking is clear only insofar as it is unconditioned by the fact that one lives the life of a human being.[11]

Other philosophers respond that these contemporary trends which reject applying concepts of humanness are themselves undesirable in that they reject too much of human experience in order to pursue goals of analytic clarity or scientific objectivity, which may not be of sufficient value to warrant the rejection of the notion of the human being in philosophical ethics.[12] In response to the challenges concerning the relevance of philosophical anthropologies in ethics one can offer to lessen the more extreme claims concerning the capacity of philosophical anthropologies to mediate or adjudicate among the various disciplines investigating human nature. Instead of claiming that philosophical anthropologies can provide an architectonic structure capable of accurately evaluating every insight from any discipline with respect to the contributions of the other disciplines, a proposition can be made to have philosophical anthropologies function through an equitable integration of the various disciplines that contribute to the understanding of human nature.[13]

The first steps toward such an integration are already taking place according to James Gustafson:

> Why do scientists draw from sources other than science when they consider what ought to be done? One answer is that they value many aspects of human life and other parts of nature. They have aesthetic appreciations and moral beliefs or aspirations—grounded only in part in their scientific work—which, to a scholar of religion, function much like myths, doctrines, and practices of religion in various human cultures. These beliefs cross the intersection of work done by theologians and other humanists. At the intersections, with prognoses of hope or fear even when confined to particular genetic interventions, both geneticists and religious thinkers are providing answers to the questions of what is valued about human life, and about particular human lives, and thus what ought and ought not to be done. The recurring systematic question is the relation of what is valued to the biologically natural or normal. This is a question for both theologians and geneticists.[14]

As Gustafson indicates, ethical analyses of human genetic interventions do, in fact, hinge upon implicit, if not explicit, philosophical anthropologies operative within the ethical frameworks used to address the issues raised by these interventions. Experts evaluating the potential uses of human genetic interventions base their conclusions on what they consider to be the crucial characteristics of human nature, if any, when deciding what should and should not be done. Often, the selection of these characteristics results from a choice, not explicitly justified, of a particular field of academic inquiry (e.g., science, philosophy, or theology) as primary or preferential in providing evidence for the characteristics selected and for the selection process itself. Thus, in some form, philosophical anthropologies are being applied to the ethical analysis of human genetic interventions. Hence, the more theoretical debate over the exact role and function of philosophical anthropologies need not be completely resolved for the purposes of this paper. Instead, a brief review of some different types of philosophical anthropologies may reveal which type might best provide a basis for an ethical evaluation of human genetic interventions.

The contemporary public discourse surrounding human genetic interventions contains many strains of thought regarding human nature. For the purposes of this paper, this diversity of thought will be grouped into four types of philosophical anthropologies: (1) static, (2) scientistic, (3) dichotomized, and (4) dynamic. Each of these four types will be evaluated as to how suitable each is to serve as a basis for the ethical analysis of human genetic interventions.

First, static philosophical anthropologies. These are based on primarily philosophical or theological beliefs about characteristics, including physical characteristics, of human nature which are considered fundamentally unchanging. Such beliefs result in the conclusion that changing these characteristics would lead to the creation of nonhuman or deficient human beings. Hence, these philosophical anthropologies proscribe any such alterations. Though obviously informed at some point in the past—even the very distant past—by biological information, these philosophical anthropologies resist integrating the findings of more recent biological research because such data may undermine their most fundamental claims about human nature.

This static approach gives precedence to philosophical and theological knowledge over scientific knowledge. Some people involved in the debates about human genetic engineering might be inclined to regard static philosophical anthropologies as advantageous since these concepts of human nature are not affected by the ever-changing results of scientific research and would, therefore, provide a more stable foundation for ethical analysis. The validity of this interpretation, however, relies on their providing a justification for giving such importance to stability and the status quo that contemporary scientific information about human nature becomes secondary or is considered subservient to the philosophical or theological concepts of humanness which undergird that moral framework.

Often in reaction to this conflict with contemporary science that can occur when employing static philosophical anthropologies, a different perspective of human nature is offered that comes from an ethical framework based solely or primarily on scientific information. Such is the second type of philosophical anthropology under evaluation here: scientistic. Scientistic philosophical anthropologies employ a kind of epistemic chauvinism similar in many respects to that employed by a static philosophical anthropology.

The principal difference between the two approaches is that each chooses a different type of knowledge as preeminent and privileged. However, as with the static type, scientistic philosophical anthropologies also require a justification for making other types of knowledge and experience, such as moral or religious, subservient to science in structuring its underlying framework. This scientific reductionism leaves no room for information from other branches of knowledge. Hence, no integration of various types of knowledge occurs and one is left with an impoverished concept of human nature rooted solely in biological or genetic determinism.

A third type of philosophical anthropology has been proposed in various forms. It presents a dichotomized interpretation of the human nature by separating human biology from moral characteristics or personhood in order to address ethical issues without having to integrate new information about human biology. This approach minimizes or rejects humanness as an important aspect in the ethical evaluation of genetic interventions. Personhood, as separate from humanness, is seen as the guiding concept toward which genetic research and intervention should be oriented.

In arguing for a dichotomistic approach, some authors go so far as to claim that using the notion of humanness as morally significant is to be guilty of a bias akin to racial prejudice. Others find no difficulty in supposing that an artificially created being, that is, a machine, could be a person. Again, it has been suggested that one's thinking is rational only insofar as it is unconditioned by the fact that one lives the life of a human being.[15]

One immediate difficulty with the dichotomized approach is that the distinction between humanness and person is not philosophically well developed. For instance, if self-conscious rationality is used as the distinguishing characteristic of personhood, then the separation of human rationality from human animality results in the estrangement of the human good from human biological flourishing. This estrangement obstructs the legitimate contributions to understanding human nature that can be made by areas of investigation such as sociobiology or medical genetics.

Additionally, whatever way one draws the line between humanness and personhood, arguments for a dichotomized approach lead to the conclusions that humanness is a completely malleable aspect of an individual's nature while personhood remains unchanging in the application of genetic technology. Such conclusions fit well with an ethical reductionism that emphasizes personal autonomy and market forces as the premier principles in moral decision making. Hence, the discoveries of research projects within medical genetics merely become the basis for providing more genetic options to be purchased by those individuals or societies who can afford them. Questions of justice and beneficence are then applied secondarily in order to assist in the adjudication of conflicts among autonomous persons. Though appearing simple, such ethical reductionism runs amok of common moral sense when individuals or groups are allowed to chose to apply genetic technologies to themselves in order to gain an advantage over others.

In contrast to philosophical anthropologies that attempt to structure and justify a separation of current scientific knowledge from other types of knowledge that inform our concepts of human nature, the last of the typologies considered here is centered around an integration of sociohistorically conditioned concepts of human nature with the findings of contemporary scientific research. These philosophical anthropologies have a dynamic character. They are not only dynamic in their constant integration of new scientific information but also in a clear acknowledgment of the fundamental dynamism of human nature.

One example of this typology is the philosophical anthropology developed by Karl Rahner, the German systematic theologian. Rahner presents human dynamism as more than just external forces shaping human nature, or humans shaping their external environment. Humans directly and fundamentally shape their own nature, and, therefore, in this sense must "manipulate" themselves.[16] Human beings achieve this self-manipulation because human nature is both radically incomplete and open to change. Change can occur because human nature has the freedom to determine what it is to become.

Rahner understands freedom and self-determination to be at the basis of human nature.[17] Freedom, for Rahner, does not mean choosing this or that, or doing this or that. Instead, it is the fundamental capacity for self-determination. Self-determination, likewise, is not simply a single free act. It is a lifelong process whereby an individual fashions himself or herself through choices and actions made in the light of that which one wants to become. Hence, freedom and self-determination involve the shaping of one's self as a whole, not merely a molding of one's body as an external vessel to meet the needs of one's rationality or genetic fitness.

This capacity for self-determination, however, is limited. It operates within parameters, constrained by social and biological limits, which act like guard rails to keep individuals and societies from taking self-destructive turns contrary to the exercise of human freedom.[18] One such parameter is human embodiment. Being embodied limits the manipulations proper to self-determination. For example, any manipulation that resulted in a significant reduction in genuine human intercommunication, even if it gave an individual a longer and less disease-ridden life, would be contrary to human freedom because it would reduce that individual's fundamental capacity for self-determination.[19]

Ultimately, though individuals exercise their freedom and shape themselves in ways that affect their entire selves, they can never exercise that freedom or shape themselves totally. Human nature remains embodied and historical, and, therefore, incomplete and contingent. No individual can fully construct himself or herself all at once at any one time, nor can human society reconstruct itself or nature to the extent that either becomes entirely a product of human manipulation. Since such complete change is not within human capabilities, it is not the direction or goal of human self-determination.

These insights led Rahner to the conclusion that human self-determination is possible only within the context of interpersonal community, since others are necessary for human development and the exercise of freedom.[20] Personhood, therefore, is also defined in terms of relation and freedom, rather than the often-used concepts of substance and reason. In contrast, then, to dichotomized philosophical anthropologies which emphasize reason in defining personhood, Rahner's approach does not allow for the removal of humanness from the moral equation because human embodiment shapes any individual's relationships and self-determination. Hence, the litmus test for genetic research and its applications is Do they serve to promote human freedom and self-determination? This standard is the culmination of Rahner's philosophical anthropology.

Reacting against static philosophical anthropologies, Rahner contends that these anthropologies

> often used a concept of "nature" ("natural," "according to nature") which ignores the fact that, although man has an essential nature which he must respect in all his dealings, man himself is a being who forms and molds his own nature through culture, i.e. in this case through self-manipulation, and he may not simply presuppose his nature as a categorical, fixed quantity.[21]

Instead, Rahner concludes that if an essence of human nature is to be discussed, then one should consider this "essence" as "not an intangible something, essentially permanent and complete, but the commission and power which enable him to be free to determine himself to his ultimate final state."[22]

Knowledge about human nature is significantly informed by contemporary scientific data. Acknowledging the need for such information, Rahner welcomed an open dialogue between science and other areas of human knowledge and research, including philosophy and theology.[23]

As noted earlier, the scientistic philosophical anthropologies also make use of contemporary scientific information in formulating their theories of human nature, but they fall short of developing frameworks which integrate that information with other types of knowledge. As a result, they do not delineate standards for the use of genetic research based on a broader scope of human thought and experience. This broader scope is particularly important because the application of genetic research will affect not just biological human nature but the whole human being.

From this sampling of four types of philosophical anthropologies, one sees different approaches to the ethical analysis of genetic research and its applications. The question, then, is this: Are all four approaches equal to the task? The answer is no.

The static philosophical anthropologies, though, perhaps, greatly concerned with preserving the value of the human being, are hindered by their selection of crucial human characteristics based on philosophical or theological tenets supported by selective, and even dated, scientific information. Consequently they cannot adequately assess contemporary scientific information without running the risk of this information contradicting what static philosophical anthropologies consider to be essential to an understanding of human nature.

Scientistic philosophical anthropologies appear to overreact in responding to the flaws in static philosophical anthropologies. In their eagerness to embrace contemporary scientific knowledge, they exclude other types of knowledge (as do the static types). This exclusion leaves these anthropologies with an impoverished picture of humanity. Reducing the human to the merely biological prevents them from taking into account the rich tapestry of human experience and interpretation. How, then, can these frameworks assess the total impact genetic research and its applications will have on human nature and society?

In order to reduce the complexity of the issues or to simplify the lines of argument from other areas of philosophy, dichotomized philosophical anthropologies seek to remove totally the biological component from the moral equation. Human biological characteristics are not considered crucial in the formulation of a philosophical anthropology. Consequently, the dichotomized frameworks cannot properly evaluate the effect changes in human bodies will have on how humans behave toward themselves and others. In the face of human experience, this denial of the importance of the human body in the moral equation requires a greater justification than its usefulness in solving certain philosophical problems.

What about dynamic philosophical anthropologies? As has been seen in the above example of Karl Rahner, dynamic philosophical anthropologies do not deny the fundamental role of scientific information in ascertaining the significant characteristics of human nature. They do not deny the importance of data and information derived from all areas of human knowledge and experience. And, finally, they do not deny the impact our bodies have on who it is human beings consider themselves to be. This openness makes it possible for the dynamic philosophical anthropologies to take into account and to evaluate the importance of all areas of human experience and, consequently, to undergird a more thorough and complete ethical analysis of scientific and technological advances.

If, then, dynamic philosophical anthropologies are to be used in the ethical analyses of human genetic interventions, what differences might there be in the various processes of ethical evaluation surrounding this issue? One immediate change would be to shift the focus away from discussions about genetic therapies versus enhancements or somatic versus germ-line interventions, because, as was argued above, these distinctions are no longer central to the issue. Instead, the focus would be on the more fundamental questions concerning human nature and human flourishing as key to human self-determination. This self-determination would involve balancing the good of society with the good of individuals, since both are necessary for the promotion of real human freedom. The articulation and understanding of real human freedom would require a broad and balanced investigation of the knowledge and wisdom humankind has acquired about it own existence and history.

To truly understand who we are and who we want to become, from the level of individuals to the level of our species, we will need to reflect on all the different fields of inquiry we have developed that inform our understanding of ourselves—such as philosophy, theology, sociology, psychology, history, literature, the fine arts, economics, law, political science, as well as the natural sciences. This broad scope of reflection will also need to be a balanced reflection. In other words, the different fields of inquiry come to the discussion as equal partners in this critical endeavor. All fields of inquiry and perspectives will have to make explicit their assumptions and presumptions, so that the critical powers of each can contribute to the overall articulation and understanding of who we are and who we want to become. Such an interaction of traditionally distinct academic subjects is already an area of intense interest for many universities and institutions. However, for this interaction to most fully inform this broad reflection, no one perspective can claim dominance over the others, especially the natural sciences.

Often, today, in public discussions about cutting-edge research we hear the claim, "We must do this because this is the best science." Pursuing a certain line of research may indeed be the best science, but that does not make it the best public policy, the best law, the best ethics, or the best for society and human self-determination. The integration of knowledge required by dynamic philosophical anthropologies aims at a more comprehensive judgment of what is best for each and all than can be found in the natural sciences alone.

Pursuit of this thorough and balanced reflection by individuals and societies does not require that all research and technological development cease until some great understanding and consensus are reached. On the contrary, this proposed process of reflection and discussion would quickly eliminate such an extreme position as counter to the intent and goal of the process itself. For example, continued research on the safety aspects of new genetic technologies, done within current internationally accepted ethical parameters, would be necessary in order that the deliberations about how to employ these new technologies reach their best possible conclusions. In the same way, efforts to bring clearly beneficial medical treatments to those in need would, at least, have to continue if not be substantially increased. The broad and balanced reflection required by dynamic philosophical anthropologies would, in fact, not entail the a priori condemnation of research into human genetic interventions. It, rather, requires the evaluation of such research within the most adequate context of human understanding.

As mentioned at the beginning of this essay, there is much media coverage and public discussion surrounding the possibilities, potentials, and perils of human genetic interventions. Intense interest in these interventions is necessary for an adequate public deliberation to take place concerning how our society and others choose to proceed with these new technologies. Intense interest is not enough, however, to guarantee an adequate public deliberation. A thorough and balanced account of our human condition will also be required so that the many and varied elements of human experience and knowledge relevant to this deliberation may be brought together in a meaningful and effective manner. This account will best be developed through the use of a dynamic philosophical anthropology, such as is present in the thought and work of the Karl Rahner.

The Rahnerian solution proposed in this essay to the problem of human genetic interventions is not the simplest nor most expedient method available for addressing this issue. It is intended instead to be thorough, just, and humane. From this perspective, then, the goal of public deliberations regarding human genetic interventions is not to see how quickly we can employ our new found abilities, but how well.

Notes

1. One interesting example of the growing global concern over public policy and genetic engineering of all types can be found in a series of commentaries done by science reporters for an issue of the journal *Nature*, which is published in England (*Nature* 403 [2000]: 5–9).

2. Though done with cellular techniques that may seem crude in comparison to some of the techniques reviewed in chap. 2, the creation of sheep/goat chimeras in the mid-1980s gives substance to this question about changing the nature of an animal by adding cells and/or genes from another animal. See C. B. Fehilly, Steen M. Willadsen, and E. M. Tucker, "Interspecific Chimaerism between Sheep and Goat," *Nature* 307 (February 16, 1984): 634–36; and C. B. Fehilly, Steen M. Willadsen, A. R. Dain, and E. M. Tucker, "Cytogenetic and Blood Group Studies of Sheep/Goat Chimaeras," *Journal of Reproduction and Fertility* 74 (May 1985): 215–21.

3. See C. Blanpain et al., "Multiple Charged and Aromatic Residues in CCR5 Amino-terminal Domain Are Involved in High Affinity Binding of Both Chemokines and HIV-1 Env Protein," *Journal of Biological Chemistry* 274 (December 3, 1999): 34719–27; C. S. Hung et al., "Relationship between Productive HIV-1 Infection of Macrophages and CCR5 Utilization," *Virology* 264 (November 25, 1999): 278–88; and L. G. Kostrikis et al., "A Polymorphism in the Regulatory Region of the CC-chemokine Receptor 5 Gene Influences Perinatal Transmission of Human Immunodeficiency Virus Type 1 to African-American Infants," *Journal of Virology* 73 (December 1999): 10264–71.

4. For example, see *Webster's Medical Desk Dictionary* (1986), s.v. "health."

5. See P. Boyle, "Shaping Priorities in Genetic Medicine," *Hastings Center Report* 25 (1995): S2–8; and William R. Clark, *The New Healers: The Promise and Problems of Molecular Medicine in the Twenty-first Century* (New York: Oxford University Press, 1997). A brief, more personal perspective offered by a clinical geneticist that can be found among other contributions is John M. Opitz, "The Geneticization of Western Civilization: Blessing or Bane?" in *Controlling Our Destinies: The Human Genome Project from Historical, Philosophical, Social, and Ethical Perspectives*, ed. Phillip R. Sloan (Notre Dame, Ind.: Notre Dame Press, 2000).

6. There are often more than two arguments listed when authors discuss why germ-line interventions should be allowed. These arguments fall into two general categories: (1) those specific to germ-line interventions, and (2) those that could be applied to other medical technologies (e.g., financial expenditures vs. savings). Since the concern of this article at this point is only with the characteristics of germ-line interventions, the arguments from the second category are not included.

7. The use of techniques that might allow for reversible changes in the DNA could meet the higher safety requirements for germ-line interventions. Two examples of such techniques are oligonucleotide gene targeting (see Seidman, M. M., "Oligonucleotide Mediated Gene Targeting in Mammalian Cells," *Current Pharmacological Biotechnology* 5, no. 5 [2004]: 321–30) and Cre-lox inserts (see Sauer, B., "Cre/lox: One More Step in the Taming of the Genome," *Endocrine* 19, no. 3 [2002]: 221–28).

8. Alex Mauron and Jean-Marie Thevoz, "Germ-line Engineering: A Few European Voices," *Journal of Medicine and Philosophy* 16 (1991): 649–66. A variation of this approach that emphasizes the need for a natural genome in order to meet the desire of the community to be able to love any new human being for him- or herself can be found in an unpublished paper by Jürgen Habermas, "On the Way to Liberal Eugenics?" The Colloquium in Law, Philosophy and Political Theory at New York University Law School (October 25 and November 1, 2001).

9. For an example of such a technology, see A. Cole-Strauss et al., "Correction of the Mutation Responsible for Sickle Cell Anemia by an RNA-DNA Oligonucleotide," *Science* (1996): 1386–89. See also an editorial on this technology by Eric Kmiec, *Gene Therapy* 6 (1999): 1–3.

10. Alison M. Jaggar and Karsten J. Struhl, "Human Nature," *Encyclopedia of Bioethics* (New York: Macmillan, 1995), 1172–73.

11. David Cockburn, ed., *Human Beings* (New York: Cambridge University Press, 1991).

12. Arguments along this line are many and varied. Analytic philosophers like David Cockburn (see *Other Human Beings* [New York: St. Martin's Press, 1990]) and William Alston ("Perceiving God," *Journal of*

Philosophy [1986]) see this rejection of the notion of the human being to involve a misguided epistemology and a false competition between explanations of physical science and everyday explanations of human relationships. Feminists such as Annette Baier emphasize the need to move away from moral theories which are pertinent only to a small group of academics ("Extending the Limits of Moral Theory," *The Journal of Philosophy* 83 [October 1986]), while sociohistorical philosophers (Charles Taylor, *Sources of the Self: The Making of the Modern Identity* [Cambridge, Mass.: Harvard University Press, 1989]) and those emphasizing praxis (Richard Bernstein, *Beyond Objectivism and Relativism: Science, Hermeneutics, and Praxis* [Philadelphia: University of Pennsylvania Press, 1983]) also point out the need for moral discourse to broaden, not rigidly limit, its scope in order to be more relevant and useful in addressing contemporary ethical concerns.

13. There are several possible frameworks from which to borrow in describing how this equitable interaction might take place. H. Tristram Engelhardt's argument for moral framework as source of communication and cooperation across boundaries and differences (*Foundations in Bioethics*, 2nd ed. [New York: Oxford University Press, 1996]) is one such. Paul Ricoeur's idea of transformative conversation as the decisive factor in *phronesis* to reach good decisions in concrete situations (*Oneself as Another*, trans. Kathleen Blamey [Chicago: University of Chicago Press, 1992]) is another possibility, as would be Richard Bernstein's attempt (see n. 12) to weave together the thought of Hans-Georg Gadamer, Jürgen Habermas, Richard Rorty, and Hannah Arendt. Or one could follow the more historical analysis of Charles Taylor in *Human Agency and Language* (New York: Cambridge University Press, 1985).

14. James Gustafson, *Intersections: Science, Theology, and Ethics* (Cleveland: Pilgrim Press, 1996), 96.

15. Cockburn, *Human Beings*, 1.

16. Karl Rahner, "The Problem of Genetic Manipulation," in *Theological Investigations* (New York: Herder & Herder, 1972), 9:227.

17. Karl Rahner, "The Experiment with Man," in *Theological Investigations* (New York: Herder & Herder, 1972), 9:213

18. Ibid., 217.

19. Karl Rahner, "The Problem of Genetic Manipulation," 233.

20. Karl Rahner, *Hearers of the Word* (New York: Seabury, 1969).

21. Karl Rahner, "The Experiment with Man," 215–16.

22. Ibid., 212.

23. Karl Rahner, *Foundations of Christian Faith: An Introduction to the Idea of Christianity*, trans. William Dych (New York: Crossroad, 1985), 8.

5

What Does Virtue Ethics Bring to Genetics?

JAMES F. KEENAN, S.J.

When ethics is allowed to be more than simply the science for determining the right and the wrong action in the here and now, it engages a variety of topics that enhances its ability to understand, describe, and evaluate practical matters. By addressing these topics, ethics becomes more competent in discerning right normative conduct. This has often not been the case.

Several contemporary writers have lamented, for instance, that bioethics has disowned its more inclusive and expansive origins in those communities of faith that first raised bioethical questions. With a tendency today to acknowledge only the secular and then only what pertains to the very specifically normative, bioethics shuts out its earlier religious companions and excludes other forms of moral discourse that engage the prophetic and the narrative.[1]

For example, Hubert Doucet compares earlier anthropological assumptions with those now operative in bioethics. He notes that when bioethics began, its respect for the person meant not simply an endorsement for personal autonomy, "but a social solidarity with those excluded from society."[2] Moreover, the first proponents of bioethics emphasized that science and technology were not simply discrete external tools for the human, but rather that *homo faber* (the human-as-maker) considered science and technology as constitutive of being human, that is, as "an integral part of the human fabric."[3] Similarly, Doucet argues that as autonomous rights developed into the overriding trump of all other claims, other constitutive elements of human self-understanding were excluded from normative discourse: "Time and space were eliminated, community and family were ignored, otherness and transcendence were made meaningless."[4]

This stripped-down self-understanding, isolated from any solidarity with society let alone with those marginalized, was given an equally stripped-down logic for making moral decisions, principalism. Its strongest advocates were Tom Beauchamp and James Childress, whose *Principles of Biomedical Ethics* offered those interested in health care four principles for moral decision-making: autonomy, justice, beneficence, and nonmaleficence.[5] The success and influence of this book on bioethics was extraordinary.

Many argued against them that they lacked an adequate appreciation of the task of ethics.[6] These arguments were so successful that twenty-two years after their first edition, the authors have recently admitted in the new fifth edition of their textbook: "We have thoroughly revised this edition of *Principles of Biomedical Ethics*, taking account of suggestions by friends and critics, developments in moral, social, and political philosophy and in biomedical ethics, and new issues in research, medicine, and health care." Among the thorough revisions is the second chapter on "moral character, especially moral virtues and ideals, which are too often neglected or downplayed in biomedical ethics."[7]

Virtue Ethics and Its Interplay with Anthropology

I am skeptical of beginning an approach to genetics that begins with principalism and simply accommodates the virtues within a "principalist" agenda. If Doucet and others are right, then much can be gained by approaching genetics not through the agenda of the rooted-in-autonomy principles but rather through the dynamic, relational virtues.

Márcio Fabri dos Anjos provides a warrant for this point of departure when he writes that good ethics "is fundamentally based on anthropological outlooks of the meaning of human life and relations."[8] Virtue ethics is that form of ethical discourse that dedicates itself to articulating human anthropological profiles so as to fill out the nature of human character. It reflects on the character traits that we ought to develop if we are to discern properly what is right or wrong about certain forms of acting in genetics. This ethics claims that we need to know and improve ourselves, if we are to know and improve the ethical agenda about what is right and wrong in genetics.[9]

Still, we need to be wary of some unexamined assumptions. Many people, for instance, tend to look at virtue ethics as if it embodies a monolith of shared, coherent substantive assumptions, but nothing could be further from the truth.[10] A comparison of the treatment of virtues by Socrates and Aristotle, for example, shows the former predominantly concerned with human reasoning, the latter giving considerable attention to friendship. Or, we could consider the preeminence that the Bible gives to the virtue of mercy, which contemporary Greek and Roman writers considered solely as a trait conveying personal ignorance rather than virtue.[11] We could read Adam Smith's *The Theory of Moral Sentiments* and find his notion of sympathy considerably at odds with what many Christians would call it.[12]

Despite these differences, virtue ethicists agree that there must be an interplay between virtue theory and anthropology.[13] The virtues are traditional heuristic guides that aim for the right realization of human identity. They are heuristic, because they are teleological, and, as such, they need to be continually realized and redefined. In a manner of speaking, their final determination and definition remain always outstanding. Being in themselves goal oriented, virtues resist classicist constructions and require us continually to understand, acquire, develop, and reformulate them. It is for this reason that inevitably Aristotle needed to develop other virtues than those that Socrates formulated.

Thus, among the many points of agreement among virtue ethicists is the first, that the virtues are dynamic. The historical dynamism of the virtues applies correspondingly to the anthropological vision of human identity which is also, in its nature, historically dynamic.[14] As we grasp better who we can become, to that extent we need to reformulate other virtues. As we determine our vision of humanity, we subsequently designate corresponding virtues to fill in or "thicken" the image of the human at which we aim.

This is not circular reasoning. We engage our understandings of both who we are as humans and who we can virtuously become through the development of history. The correspondence between the two is, therefore, more akin to a spiral than to a circle, for the interplay between our anthropological self-understanding and our virtuous goals advances through history.[15]

The second point of agreement is that the virtues always point us toward a social, relational self-understanding. This is because, being concerned with anthropology, we cannot imagine the individual human as the only member of the species. The human being always belongs to humanity. Even a virtue ethics rooted in the monastic or the hermitic tradition still discusses the person as a member or participant of an on-going tradition.

Third, inevitably, because of its social and dynamic orientation, virtue privileges justice and aims at establishing equality. We see this goal emerge, for example, by reflecting first on Aristotle's elitism, which led him to discuss virtues primarily for those who could be magnificent. While he designated other virtues for educated men, he did not develop any per se for women or slaves. But as philosophers and theologians further developed a more inclusive understanding of justice, they needed other virtues that substantiated a more democratic framework.

Though the virtues aim for equality in the name of justice, a common misunderstanding emerges in assuming that virtue ethicists are strictly communitarian and sectarian. Certainly some virtue ethicists, among them Alasdair MacIntyre, insist on this model.[16] Others like Martha Nussbaum, however, rightly recognize that these sectarian, communitarian concerns are not intrinsic to virtue theory.[17]

Here then emerges a fourth point that many ethicists agree on: the potential compatibility of virtue theory with liberal societies. This claim is particularly important since from a variety of locations we notice a shift in anthropological assumptions that virtue theory can considerably support. For instance, the moral theology coming out of western Europe basically maintains its original agenda as an autonomous ethics in the context of faith, that is, an ethics of the conscience inspired by God. As a basic telos, however, autonomy or conscience offers in itself no adequate expression for the end of the human subject. Thus most European theologians insist on the need to talk of the realization of an autonomous subject as constitutively relational. A profound interest in the person who is constituted by solidarity with others (neighbor, God, and nature) is both the anthropological given and the moral task.[18] The solidarity that virtue ethics provides answers the need of liberal societies to protect the individual conscience while at the same time protecting the subject from the isolating, stripped-down anthropology and its attendant prinicipalism that has hampered the development of liberal cultures.

One particular European model is the personalism prevalent throughout northern Europe. Paul Schotsmans, for instance, recognizes that the significance of the individual person as unique but places that insight in tension with the theological fact that every human creature is constitutively related to the creator. Schotsmans frames the fundamental question for normative conduct in every instance: "how can the personal decisions of conscience of a unique person express and integrate this connection with a solidary responsibility for society?"[19]

Similarly in the United States, many moral theologians recognize the importance of understanding the centrality of the individual acting in conscience, but always as a relational person called in solidarity with others.[20] For instance, Lisa Cahill's essay in this collection, as well as many of her other writings, describes how the engagement of feminism with a natural law ethics based on the common good can promote justice and solidarity in a contemporary liberal world.[21]

The Need to Propose Specific Cardinal Virtues for Contemporary Anthropological Claims

Though the virtues help us to appreciate the need to foster a constitutively relational anthropology that promotes equality and solidarity while protecting the individual conscience, they also help us to appreciate that our anthropology is historically emerging. Because we are seeking a telos that remains on the horizon of our expectations and therefore because our virtues are, as we have seen, heuristic guides, we need also to recognize the fact that the virtues could be in competition with one another.

William Spohn, for instance, contends that contemporary virtue ethicists generally presume that virtues conflict.[22] They share with other ethicists the presupposition, then, that conflict among key directing guidelines is inherent to all contemporary methods of moral reasoning. For instance, William Frankena, after presenting the two fundamental principles of beneficence and justice, raises "the problem of possible conflict" between the two principles and writes, "I see no way out of this. It does seem to me that the two principles may come into conflict, both at the level of individual action and at that of social policy, and I know of no formula that will always tell us how to solve such conflicts."[23] Likewise Tom Beauchamp and James Childress argue that "there is no premier and overriding authority in either the patient or the physician and no preeminent principle in biomedical ethics—not even the admonition to act in the patient's best interest."[24]

On the supposition of possible conflict, I have proposed a contemporary view of the cardinal virtues that corresponds to this contemporary understanding of anthropology.[25] I suggest that our identity is relational in three ways: generally, specifically, and uniquely. Each of these relational ways of being demands a cardinal virtue: as a relational being in general, we are called to justice; as a relational being specifically, we are called to fidelity; as a relational being uniquely, we are called to self-care. These three virtues are cardinal. Unlike the classical foursome, none is ethically prior to the other. They have equally urgent claims and they should be pursued as ends in themselves: we are not called to be faithful and self-caring in order to be just, nor are we called to be self-caring and just in order to be faithful. None is auxiliary to the others. They are distinctive virtues with none being a subset or subcategory of the other. They are cardinal. The fourth cardinal virtue is prudence, which determines what constitutes the just, faithful, and self-caring way of life.

Our relationality generally is always directed by an ordered appreciation for the common good in which we treat all people as equal. As members of the human race, we are expected to respond to all members in general equally and impartially.[26] If justice urges us to treat all people equally, then fidelity makes distinctively different claims. Fidelity is the virtue that nurtures and sustains the bonds of those special relationships that humans enjoy whether by blood, marriage, love, citizenship, or sacrament. If justice rests on impartiality and universality, then fidelity rests on partiality and particularity. Neither of these virtues, however, addresses the unique relationship that each person has with oneself. Care for self enjoys a considered role in our tradition, as, for instance, the command to love God and one's neighbor as oneself. In his writings on the order of charity, Thomas Aquinas, among others, developed this love at length.[27]

Finally, prudence has the task of integrating the three virtues into our relationships. Thus, prudence is always vigilant, looking to the future, trying not only to realize the claims of justice, fidelity, and self-care in the here and now but also calling us to anticipate occasions when each of these virtues can be more fully acquired. In this way prudence is clearly a virtue that pursues ends and effectively establishes the moral agenda for the person growing in these virtues. But these ends are not in opposition to nor in isolation from one another. Rather prudence helps each virtue to shape its end as more inclusive of the other two, that is, the conflict among the virtuous claims are to be eventually integrated.

Prudence does not work alone. Aristotle departed from Socrates' belief that prudence is sufficient for self-realization and self-determination. Prudence, Aristotle warned us, depended on the other virtues, and those virtues were dialectally dependent on prudence.[28] For this reason, the competency of prudence is deeply imbedded in the historicity of human beings such that human beings can only perceive well the horizon of their possibilities to the extent that they have rightly realized themselves through the virtues.[29] The spiral engages human nature, human vision, and human reasoning.

This prudence is worked out then in the narrative of the lives we live. Stanley Hauerwas reminds us that we have the task of sorting out "conflicting loyalties" throughout our lives. That sorting out means that we must incorporate the variety of relational claims being made on us: we do this through the narrative of the lives we live.[30] The virtues are related, then, to one another not in some inherent way as they seem to be in the classical list of the cardinal virtues. The virtues do not per se complement one another; rather, they become integrated in the life of the prudent person who lives them. The unity of the virtues is found not in the theoretical positing of the cardinal virtues, but rather in the final living out of lives shaped by prudence anticipating and responding to virtuous claims.[31]

Paul Lauritzen has done an important synthesis of recent works on morality and the self in which he argues that the turn to narrative ethics has enabled us to see both that the self as fragmented becomes integrated by the narrative one lives and that, as he writes, "the narrative self is necessarily a social and relational self." A relational anthropology requires us then to rethink our understanding not only of the self but of morality in general and of the cardinal virtues in particular.[32]

It is hard to imagine any culture, therefore, that would not minimally recognize the claims of partiality and impartiality as well as some sense of self-regard. I suggest that all persons in every culture are constituted at least minimally by these three ways of being related. Moreover, by naming these virtues as cardinal, we have a device for talking cross-culturally. This device is based, however, on modest claims. The cardinal virtues do not purport to offer a picture of the ideal person, nor to exhaust the entire domain of virtue. Rather than being the last word on virtue, they are among the first, providing the bare essentials for right human living and specific action. As hinges, the cardinal virtues provide a skeleton of both what human persons should basically be and at what human action should basically aim. All other issues of virtue hang on the skeletal structures of both rightly integrated dispositions and right moral action.

In particular cultures, therefore, these virtues need to be thickened. Cultures give flesh to the skeletal cardinal virtues. This thickening differentiates one virtue in one culture from a similar one in another. While we can differentiate in each culture the virtue that aims at impartiality from that which promotes partiality, we can likewise compare and contrast those specific virtues in the different cultures. Justice, fidelity and self-care in a Buddhist culture have somewhat similar and somewhat different meanings than they do in a liberal or a Confucian context.[33]

Christian Catholic cultures also thicken these virtues, usually through mercy, a virtue that has a certain priority in Christianity. I cannot in the space here develop that claim,[34] except to say that in Catholic cultures mercy "thickens" our understanding of justice, fidelity, and self-care. Inasmuch as mercy is the willingness to enter into the chaos of another so as to respond to the other, justice thickened by mercy insists on taking into account the chaos of the most marginalized. Fidelity in a Catholic marriage is thickened by the yeast of mercy, and therefore spouses enter the other's chaos seventy times seven times. Finally a Christian virtue of self-care urges the self to enter into the deep chaos of one's own distinctively complicated life.

Advancing Ethical Discourse on Genetics through Virtue Ethics

When we bring the virtues into the question of genetics, a variety of issues emerges. First, the teleological structure of virtue ethics makes it a worthy companion to genetic research. Because it is so dynamic yet tentative, because it depends upon our on-going understanding of human capabilities, and because it possesses within its own structure of reflection the sense of a continuously emergent ethics, virtue ethics tracks a path of reflection that is not unlike the framework of genetic research itself. Both appreciate the fact that data and judgment go hand-in-hand and the double helix is as much of a spiral as is the hermeneutical context for virtue ethics going forward.

This compatibility is not, however, because virtue ethics is simply fluid in its ability to proffer guidance. Often virtue ethics is dismissed out of hand because its critics argue that it fails to give teeth to the moral insights that it proffers.[35] This, however, is more the abuse of virtue ethics than the claims of virtue ethics.

For instance, recently I examined the use of virtue language as it appeared in the papers commissioned by the National Institutes of Health to study embryonic research.[36] Those who insisted on the legitimacy of their research invoked a "respect" for human embryonic life as giving their claims moral credibility. They usually held that they would not subject embryonic life to "demeaning" research, for example, for cosmetic purposes.[37] John Robertson claimed to have a similarly respectful disposition and described what is at stake in this respect by noting that the entire issue is "about what cost in foregone knowledge should be tolerated to demonstrate the respect for human life that limiting embryo research symbolizes."[38] For Robertson and others, respect is the price we pay for research. Their respectful dispositions were not based on generating any positive guidelines on how to treat embryonic life, rather they were used as a political "wedge" to keep them from doing "demeaning" research. That human embryos could be mass produced, cloned, manipulated, and discarded was not subject to this "respect."

About the use of respect, then, Courtney Campbell raised concern about "how the moral respect the panel confers on the embryo will be meaningful in the context of scientific research."[39] Daniel Callahan wrote, "it is better and more honestly done by simply stripping preimplantation embryos of any value at all. If we look under the rhetoric of respect, that seems to me the actual meaning of what the panel has done. At best, the kind of respect it would accord embryos is to them as a class, not as individual embryos. Those embryos that stand in the way of research are to be sacrificed—as nice a case of the ends justifying the means as can be found."[40]

The rhetoric of virtue then is subject to the same critique as the rhetoric of value or responsibility. Though some may attempt to employ a virtue to camouflage or mute the concerns of others, for the most part, a virtue must demonstrate and defend its credibility in action. The virtue of respect is not truly operative in Robertson's claims. We can use virtues only if we expect them to offer us some real concrete positive guidelines toward becoming the type of people we should become.

Second, because of its teleological nature, virtue ethics has a natural tendency to be open to the distant horizons of genetic research. Generally, it is not inclined to preempt a particular type of genetic manipulation, for example, germ-line gene therapy or genetic enhancement. For instance, several European ethicists and associations have argued against germ-line gene therapy by declaring the right to inherit a genetic pattern that has not been artificially changed.[41] This assertion, as far as I understand it, places an insurmountable obstacle against germ-line gene therapy. Virtue ethicists would argue otherwise. They would claim that germ-line gene therapy could prompt many goods. They would begin their argument by noting that parents and health-care personnel have to develop dispositions to prudently and faithfully act on behalf of the interests of children.[42] Are not these virtuous dispositions precisely the guarantees for altering the genes of our progeny that "nature" randomly gave them?

That openness is countered, fortunately, by a third way that virtue ethics specifically enhances the ethical study of genetic research. Because virtue ethics examines the capabilities of human agents in the here and now, virtue ethicists are not naive optimists but rather acknowledge the "imprudence" of validating certain procedures in the here and now. For instance, in light of the fact that so few researchers actually demonstrate any evident concern for embryonic life today, many virtue ethicists welcome strict guidelines imposed on researchers who work with early forms of human embryonic life. Similarly, while these ethicists might be inclined to entertain questions about genetic enhancement, they would at the same time examine how inclined the human community is to prudentially weigh concerns about justice and fairness in the face of such enhancement.[43]

Virtue ethicists recognize the possibilities for the future. Generally they oppose measures that could preclude forever future opportunities. At the same time, they are critically aware of the limitedness of our ability to be just, faithful, self-caring, and prudential in the present management of these responsibilities. This tension between recognizing both what we will be capable of doing and what our present moral dispositions actually are is a tension that virtue ethics brings to the table of reflection. Indeed, many today have considerable suspicion of the motives operative in much embryonic and genetic research, and virtue ethics's specific interest in analyzing the dispositions that underlie moral action provides a method for critically reviewing those dispositions.[44]

Fourth, virtue ethicists explore how moral action inevitably affects the moral agent. Following Aristotle and Aquinas, we use phrases like "we become what we do."[45] Virtue ethicists therefore remind geneticists that their work is not a manipulation of some object-out-there but inevitably is a manipulation of our very selves.[46] Genetic research eventually redounds to our own human nature: by it, we are basically affecting humanity itself. Not only are we affecting our progeny but because they are our progeny we are affecting our human race. Thus virtue ethics naturally insists on the participation of the rest of humanity in any dialogue on genetics because of their vested (self-)interest in the ethics of genetic research and specifically in the work of geneticists.

Fifth, the conflict among virtues which parallels the conflict among goods is particularly helpful as we try to recognize and differentiate the different loyalties at play in genetics. In fact, I suggest that the virtues proposed earlier help us to recognize and eventually arbitrate the various conflicting claims at stake in a variety of debates about genetics.

Consider, for instance, policy-making debates over the use of embryonic stem cells for therapeutic purposes. On the one hand, we recognize exponents of the virtue of justice who are disposed to consider to what extent all similar forms of human life ought to be similarly protected. Here we see frequently the interest of impartiality advanced as religious leaders insist that defenseless human life anywhere must be defenseless human life protected everywhere.[47]

On the other hand, many participants in the discussion do not hesitate to reveal that their interests are specifically about their own health prognosis (self-care) or about particular family members or those others who share their own or their relatives's prognosis (fidelity). That is, many do not hesitate to argue that their ethical disposition to care for their own health or for the health of those to whom they claim some sort of particular relationship requires them to advocate for this very partial interest.

Some ethicists, who only recognize the claims of justice, overlook or actually dismiss these more fiduciary concerns. Others, like Paul Ricoeur, recognize the inadequacy of justice and impartiality as the exclusive arbiter of ethically right determinations: the bonds of partiality also enter into ethical reflection.[48]

Prudence needs then to arbitrate both the claims of equality that pertain to human life (especially to the extent that human embryonic life participates in those claims) and the claims of fidelity and self-care by which we advocate on behalf of those whose health prognosis could possibly improve by the use of embryonic stem cell tissue.[49] Margaret Farley, for instance, develops the tension between partiality and impartiality[50] and brings these two into play as she addresses the discussion of embryonic stem cells.[51]

Sixth, while the virtue of prudence prompts us to arbitrate the conflicting dispositions that dominate much contemporary discourse, it also has the long-range role of eventually reconciling the divergent dispositions of justice, fidelity, and self-care. Here, then, we see that prudence cannot simply resolve conflicts for the moment but must create a way of proceeding that anticipates ways of diminishing the gulf between dispositions of impartiality and partiality.

Here, then, we see that virtue ethics, again as teleologically oriented, recognizes the need to continually work in the present to better our ability to act for the future. In responding then to contemporary needs, prudence reflects on how a proposed resolution for a particular conflict actually incorporates ways for anticipating and actually resolving conflicts for the future.

Here we can think of the many issues regarding genetics that stem from the industrialized world's own interests in caring for itself (self-care) and for its kin (fidelity). Often, however, those concerns overlook the urgent medical questions for all of humanity (justice) where equal access to health care has more rudimentary and less technologically oriented interests.[52]

Prudence calls us then to construct responses now to a variety of issues related to genetics that will make us more disposed in the future to respond with an even greater sense of fairness and fidelity. Looking to the future, virtue ethics challenges us to anticipate in today's actions a way of becoming better disposed to arbitrate those that lay on the horizon of our expectations.

Finally, virtue ethics always reminds us that when we engage in moral decisions we are inevitably engaging in an understanding of our humanity. Unlike the principalists and others, virtue ethics very intentionally brings to the table of discussion both our humanity and the variety of ways that we are humanly related. It continually prompts us to recognize that as we act we are shaping ourselves and therefore it asks us to define who the "ourselves" really are. If some persons only favor a partial understanding of ourselves, or even a completely self-oriented understanding of ourselves, then others can rightly insist that until the three claims of justice, fidelity, and self-care are prudentially brought to the table of decision, the discussion and the subsequent decisions are, like the agents themselves, not even minimally acting from a moral context.

Notes

1. See *Theology and Bioethics,* ed. Earl E. Shelp (Dordrecht: D. Reidel, 1985); Charles Curran, "Moral Theology and Dialogue with Biomedicine and Bioethics," *Studia Moralia* 23 (1985): 57–59; Stephen Lammers, "The Marginalization of Religious Voices in Bioethics," in *Religion and Medical Ethics: Looking Back, Looking Forward,* ed. Allen Verhey (Grand Rapids: Eerdmans, 1996), 19-43; see Lisa Sowle Cahill's essay in this collection and in particular her reference to Daniel Callahan, "The Social Sciences and the Task of Bioethics," *Daedalus* 128 (1999), 275-94.

2. Hubert Doucet, "How Theology Could Contribute to the Redemption of Bioethics from an Individualist Approach to an Anthropological Sensitivity," *The Catholic Theological Society of America's Proceedings* 53 (1998): 53–66, here 54. Doucet borrows from Mark E. Meany, "Freedom and Democracy in Health Care Ethics: Is the Cart before the Horse?" *Theoretical Medicine* 17 (1996): 399–414.

3. Doucet, "How Theology Could Contribute to the Redemption of Bioethics," 55; see P. S. Greenspan, "Free Will and the Genome Project," *Philosophy and Public Affairs* 22 (1993): 31–43; Gerald P. McKenny, *To Relieve the Human Condition* (Albany: State University of New York, 1997).

4. Doucet, "How Theology Could Contribute to the Redemption of Bioethics," 55.

5. Tom Beauchamp and James Childress, *Principles of Biomedical Ethics* (New York: Oxford University Press, 1979).

6. The debates were extensive. See, for instance, Edwin DuBose, Ron Hamel, and Laurence O'Connell, eds., *A Matter of Principles* (Valley Forge, Pa.: Trinity Press International, 1994); Peter A. French et al., eds., *Midwest Studies in Philosophy 13: Ethical Theory: Character and Virtue* (Notre Dame, Ind.: University of Notre Dame Press, 1988).

7. Beauchamp and Childress, *Principles of Biomedical Ethics,* vii.

8. Márcio Fabri dos Anjos, "Medical Ethics in the Developing World: A Liberation Theology Perspective," *Journal of Medicine and Philosophy* 21 (1996): 635.

9. See James Walter's claim in "A Response to Hubert Doucet," *The Catholic Theological Society of America's Proceedings* 53 (1998): 67–71. See also Walter, "Perspectives on Medical Ethics: Biotechnology and Genetic Medicine," in *A Call to Fidelity: On the Moral Theology of Charles E. Curran,* ed. James J. Walter, Timothy O'Connell, and Thomas Shannon (Washington, D.C.: Georgetown University Press, 2002): 135–52.

10. Helpful here is Alasdair MacIntyre's insight in the second half of his work *After Virtue: A Study in Moral Theory* (Notre Dame, Ind.: University of Notre Dame Press, 1981), 114–245

11. Rodney Stark, *The Rise of Christianity: A Sociologist Reconsiders History* (Princeton, N.J.: Princeton University Press, 1996).

12. Martin J. Calkins and Patricia H. Werhane, "Adam Smith, Aristotle, and the Virtues of Commerce," *Journal of Value Inquiry* 32 (1998): 43–60.

13. I have developed these thoughts in "Virtue and Identity," in *Creating Identity: Biographical, Moral, Religious,* ed. Hermann Häring, Maureen Junker-Kenny, and Dietmar Mieth (*Concilium* 2000/2; London: SCM Press, 2000): 69–77.

14. Michael Himes, "The Human Person in Contemporary Theology: From Human Nature to Authentic Subjectivity," in *Introduction to Chris-*

tian Ethics: A Reader, ed. Ronald Hamel and Kenneth Himes (Mahwah, N.J.: Paulist Press, 1989), 49–62.

15. See this same hermeutical engagement in William Schweiker, *Power, Value and Conviction: Theological Ethics in the Postmodern Age* (Cleveland: Pilgrim Press, 1998); and Thomas Kopfensteiner, "Science, Metaphor and Moral Casuistry," in *The Context of Casuistry,* ed. James Keenan and Thomas Shannon (Washington, D.C.: Georgetown University Press, 1995), 207–20; Thomas R. Kopfensteiner, "The Metaphorical Structure of Normativity," *Theological Studies* 58 (1997): 331–46.

16. Alasdair MacIntyre, *Whose Justice? Which Rationality?* (Notre Dame, Ind.: University of Notre Dame Press, 1988).

17. Martha Nussbaum, "Non-Relative Virtues: An Aristotelian Approach," *Midwest Studies in Philosophy 13,* ed. Peter A. French et al., 32–53. See also David Solomon, "Internal Objections to Virtue Ethics," *Midwest Studies in Philosophy 13,* 428–41.

18. James F. Keenan and Thomas Kopfensteiner, "Moral Theology out of Western Europe," *Theological Studies* 59 (1998): 107–35. See, for instance, Dietmar Mieth, ed., *Moraltheologie im Abseits? Antwort auf die Enzyklika "Veritatis Splendor"* (Freiburg: Herder, 1994).

19. Paul Schotsmans, "In Vitro Fertilisation: The Ethics of Illicitness? A Personalist Catholic Approach," *European Journal of Obstetrics and Gynecology and Reproductive Biology* (1998): 4.

20. Stephen Pope, "The Order of Love and Recent Catholic Ethics: A Constructive Proposal," *Theological Studies* 52 (1991): 255–88. Similarly, Augustine writes, "the life of the wise man should be social" (*City of God* 19.5).

21. Similarly, see Christina Traina, *Feminist Ethics and Natural Law* (Washington, D.C.: Georgetown University Press, 1999).

22. William Spohn, "The Return of Virtue Ethics," *Theological Studies* 53 (1992): 60–75.

23. William Frankena, *Ethics,* 2nd ed. (Englewood Cliffs, N.J.: Prentice-Hall, 1973), 52.

24. Beauchamp and Childress, *Principles of Biomedical Ethics,* 211.

25. James Keenan, "Proposing Cardinal Virtues," *Theological Studies* 56, no. 4 (1995): 709–29.

26. See Paul Ricoeur, "Love and Justice," in *Radical Pluralism and Truth: David Tracy and the Hermeneutics of Religion,* ed. Werner G. Jeanrond and Jennifer L. Rike (New York: Crossroad, 1991), 187–202.

27. Stephen Pope, "Expressive Individualism and True Self-Love: A Thomistic Perspective," *Journal of Religion* 71, no. 3 (1991): 384–99; Edward Vacek, *Love, Human and Divine* (Washington, D.C.: Georgetown University Press, 1995), 239–73.

28. Aristotle, *Nicomachean Ethics*, VI.1144b10–1145a11.

29. Daniel Mark Nelson, *The Priority of Prudence: Virtue and the*

Natural Law in Thomas Aquinas and the Implications for Modern Ethics (University Park, Pa.: Pennsylvania State University Press, 1992).

30. Stanley Hauerwas, *A Community of Character* (Notre Dame, Ind.: University of Notre Dame Press, 1981), 144.

31. On the task of integration, see Gedaliahu Stroumsa, "*Caro salutis cardo*: Shaping the Person in Early Christian Thought," *History of Religions* 30 (1990): 25–50.

32. Paul Lauritzen, "The Self and Its Discontents," *Journal of Religious Ethics* 22, no. 1 (1994): 189–210, here 206. See also Dietmar Mieth, "Moral Identity—How Is It Narrated?," in *Creating Identity,* 11–22.

33. Lee H. Yearley, *Mencius and Aquinas: Theories of Virtue and Conceptions of Courage* (Albany: State University of New York Press, 1990).

34. James Keenan, "Mercy: What Makes Catholic Morality Distinctive," *Church* 16 (2000): 41–43; "The Works of Mercy," *Church* 16, no. 4 (2000) 39–41.

35. This is a common mistake which David Solomon examines in his "Internal Objections to Virtue Ethics," in *Midwest Studies in Philosophy 13*, ed. Peter A. French et al., 428–41. As critic, see Sarah Conly, "Flourishing and the Failure of the Ethics of Virtue," in ibid., 83–96.

36. James F. Keenan, "Casuistry, Virtues, and the Slippery Slope: Major Problems with Producing Human Embryonic Life for Research Purposes," in *Cloning and the Future of Human Embryo Research,* ed. Paul Lauritzen (New York: Oxford University Press, 2000), 67–81. See also Carol Tauer, "Responsibility and Regulation: Reproductive Technologies, Cloning, and Embryo Research," in ibid., 145–61.

37. Bonnie Steinbock, ed., *Papers Commissioned for the NIH Human Embryo Research Panel* (Bethesda, Md.: National Institutes of Health, 1994). Similarly, see Dena Davis, "Embryos Created for Research Purposes," *Kennedy Institute of Ethics Journal* 5 (1995): 343–54.

38. John Robertson, "Symbolic Issues in Embryo Research," *Hastings Center Report* 25, no. 1 (1995): 37–38, here 38.

39. Courtney Campbell, "Response to Hogan and Green," *Hastings Center Report* 25, no. 3 (1995): 5. Campbell was "troubled by doubt as to whether the philosophical value the panel has attributed to the preimplantation human embryo will have any moral significance within the research process." See Courtney Campbell, "Awe Diminished," *Hastings Center Report* 25, no. 1 (1995): 44–46, here 44.

40. Callahan, "The Puzzle of Profound Respect," *Hastings Center Report* 25, no. 1 (1995): 39-40, at 40.

41. Maurice de Wachter, "Ethical Aspects of Human Germ-line Gene Therapy," *Bioethics* 7 (1993): 166–77; A. Mauron and J. Thevoz, "Germline Engineering: A Few European Voices," *Journal of Medicine and Philosophy* 16 (1991): 667–84.

42. This type of argument is found in John Fletcher, "The Brink: The Parent-Child Bond in the Genetic Revolution," *Theological Studies* 33 (1972): 457-85; John Fletcher and Dorothy Wertz, "Privacy and Disclosure in Medical Genetics Examined in an Ethics of Care," *Bioethics* 5 (1991): 212–27; Kathleen Nolan, "Commentary: How Do We Think about the Ethics of Human Germ-line Genetic Therapy?" *Journal of Medicine and Philosophy* 16 (1991): 613–20; eadem, "First Fruits: Genetic Screening," *Hastings Center Report* 22, no. 4 (1992): S2. See also James Keenan, "Genetic Research and the Elusive Body," in *Embodiment, Medicine and Morality,* ed. Margaret Farley and Lisa Sowle Cahill (Dordrecht: Kluwer Academics, 1995), 59–73.

43. James Keenan, "Whose Perfection Is It Anyway?: A Virtuous Consideration of Enhancement," *Christian Bioethics* 5 (1999): 104–20; Gerald McKenny, "Enhancements and the Ethical Significance of Vulnerability," in *Enhancing Human Traits: Conceptual Complexities and Ethical Implications*, ed. Erik Parens (Washington, D.C.: Georgetown University Press, 1999); Erik Parens, "Is Better Always Good? The Enhancement Project," *Hastings Center Report* (January-February 1998): S1–115.

44. See some of the testimonies in *Ethical Issues in Human Stem Cell Research, Religious Perspectives,*" National Bioethics Advisory Commission," June 2000.

45. Thomas Aquinas noted that virtue is about immanent as opposed to transient actions, that is, virtue is about actions that we do as opposed to actions by which we make things. Thomas Aquinas, *Summa Theologiae* I–II 57.5 ad 1; 68.4 ad 1.

46. James Keenan, "What Is Morally New in Genetic Engineering?" *Human Gene Therapy* 1 (1990): 289–98.

47. See, for instance, Edmund Pellegrino's "Testimony," in *Ethical Issues in Human Stem Cell Research,* F 1–3.

48. Ricoeur, "Love and Justice."

49. Certainly we need also to recognize that many who advocate for embryonic research for therapeutic purposes also invoke impartial justice as grounds for their claims.

50. Margaret Farley, "An Ethic for Same Sex Relations," in *A Challenge to Love,* ed. Robert Nugent (New York: Crossroad, 1986), 93–106; and eadem, *Personal Commitments: Beginning, Keeping, Changing* (San Francisco: Harper & Row, 1990).

51. Margaret Farley, "Testimony," in *Ethical Issues in Human Stem Cell Research, Religious Perspectives*, D1–5.

52. Márcio Fabri dos Anjos's and Lisa Sowle Cahill's essays in this volume address these concerns. See also Cahill, "Stem Cells: A Bioethical Balancing Act," *America* (March 26, 2001): 14–19; and eadem, "The Genome Project," *America* (August 12, 2000): 7–13.

PART I

PROJECT CONTRIBUTIONS

Section III: Socioeconomic Significance

6

Genetics, Theology, Common Good[1]

LISA SOWLE CAHILL

In 2000, leaders of the international Human Genome Project announced jointly with a private U.S. corporation, Celera, that the human genome had been almost completely sequenced. Exhilaration over this impressive accomplishment was enhanced by anticipation of potential benefits to human health, which also spurred a surge in biotech and pharmaceutical investments, as well as gene-related patent applications. Celera's race to decode the genome before the HGP illustrates that patent rights over genetic knowledge will offer big business opportunities for entrepreneurs of the twenty-first century.

The social and ethical implications of genetic research cannot be understood apart from economic globalization, worldwide health inequities, and the absence of any one institution or well-coordinated set of institutions that can define and govern a "world order."[2] Much bioethical discussion of genomics, however, centers not on fair access and distribution of burdens and benefits globally but on the privacy, choice, and rights of scientists, investors, and clients in the so-called developed countries, where most people have access to at least minimal medical care. The individualism of researchers, corporations, clinicians, and consumers in the first world has already constructed a debate about the social ethics of human genetics that centers on balancing the interests of those fortunate enough to buy and sell in the national or global biotech marketplace. The current "pro-commercialization environment"[3] for the deployment of biotech advances results from the combination of a consumer culture, the absolutization of autonomy as a social and political value, and market forces that permeate the biotechnology sector.

The right of an individual to participate in the market economy by selecting and consuming products has become an important "emblem of individual freedom," working against uniform and effective international laws limiting or banning even the creation of research embryos or human cloning. Governments perceived as "anti-biotech" may drive research and investments into another region or country.[4] The primacy of financial assets and choice is likely to be just as great an influence on future genetic innovations, such as genetically targeted drugs, tissue replacements derived from stem cells, and genetic enhancements. Market pressures help define consumer perception of need, both directly and indirectly, abetted by media that follow genetic developments avidly, sensationalize discoveries, and polarize supporters and opponents of the latest innovation. Public debate about genetic developments concentrates mainly on the moral status of embryos used in research and the possibility that individual rights might be infringed if the confidentiality of genetic tests is not secure. In Europe more than in the United States, there is also fear that genetic knowledge and power could unleash a new socially discriminatory eugenics movement. Concerns such as these are weighed in the balance against the supposedly great promise of genetics to relieve human suffering and improve the human condition. Advocacy for genomics as a major medical advance is promoted by researchers and businesses that have a significant financial interest in keeping regulations and restraints at a minimum. Although investment in the private sector may provide a powerful impetus to research that produces benefits, this positive result should be assessed within a social and moral framework that also includes distributive justice and access to health care.

As has been observed in the introduction to the present collection, a comprehensive and just approach to genetics and ethics can benefit from the insights of religious traditions and theology about the importance of an inclusive vision of the common good, guided in particular by solidarity and a preferential option for the poor. Key elements of the Catholic common good tradition will be developed further below. First it is important to note that recognition of the importance of balancing individual and communal interests in pursuing new types of genetic research is hardly absent even at the level of international bioethics policy.

International Policy and the Common Good

References to human interdependence and the requirements of cooperative living may be found at the international level, particularly in statements influenced by European perspectives. The UNESCO Universal Declaration on the Human Genome and Human Rights (1997) recalls that the preamble of UNESCO's constitution refers to the dignity and mutual respect of all persons, enjoins on nations "a spirit of mutual assistance and concern," proclaims that "peace must be founded on the intellectual and moral solidarity of mankind," and defines "the common welfare of mankind" as an objective of the United Nations.[5] Exactly how these admirable common goals are to be realized is more difficult to establish, especially since not all those who subscribe to them are willing to risk personal or national advantage for the sake of the general welfare. Hence, the declaration allows patenting and leaves the door open to some forms of human cloning ("nonreproductive" forms). Yet even though the declaration acknowledges international law protecting intellectual property, it emphasizes the "biological diversity of humanity" and affirms the inherent dignity and "equal and inalienable rights of all members of the human family." This dignity and these rights should be respected in the course of genetic research, with its prospects for "improving the health of individuals and of humankind as a whole." Individuals, families, and groups who are vulnerable to genetic disadvantage should be respected through a "practice of solidarity" promoted by the state.[6]

In an assertion whose implications have been debated, the declaration also states in article 1 that, "The human genome underlies the fundamental unity of all members of the human family, as well as the recognition of their inherent dignity and diversity. In a symbolic sense, it is the heritage of humanity."[7] Christian Byk, of the International Association of Law, Ethics, and Science (Paris), sees the inspiration of this statement in a philosophy of universal human rights, and observes that it marks the first time that "humanity, the human species, has a common heritage to be protected."[8] One way to interpret the claim that the genome is the common heritage of all and the property of none is through a utilitarian perspective that would deny that individual rights or autonomy can override what is decided (by whom?) to be in the best genetic interests of the whole of humanity or of its "gene pool." Such an interpretation would tend to justify the sort of eugenic programs that had such disastrous consequences only a generation ago in Europe and which are reflected today in equally brutal programs of "ethnic cleansing" or its equivalent around the globe. This is a particular concern of some German authors.[9] This, however, is not the interpretation favored by Byk. Instead, the definition of the genome as a "common heritage" equates to the moral duty to protect all members of society, both living and future, from abuse, and to confront the fact that we could all be perpetrators as well as victims.

Byk further observes that the declaration does contain a tension between the idea that the genome is a "common heritage" and the duty to protect individual rights. It is not clear which should prevail in cases of conflict. Nor is it clear who has the responsibility to protect either the genome as common heritage or the subjective rights of the individual. Without resolving these problems, Byk offers two principles for beginning to address them. First, the obligation to protect the genome belongs to society as a whole, not just to researchers or to science. The application of research in biology is a political, cultural, and social issue, and the public must or should participate in the debate on the basis of accurate information and a democratic process. Second, the declaration adopts a positive attitude toward research and its benefits but provides that such benefits "shall be made available to all" (article 12). In other words, no individual shall be excluded from full participation in the welfare that is sought in common. In effect, this provision also extends the rights of individuals beyond access to basic health care to access to new genetic interventions.[10] Unfortunately, however, the declaration does not shed much light on the difficult task of specifying procedures, regulations, or structures for guaranteeing individual rights or ensuring that the genome will be respected and explored for the benefit of humankind in general. Indeed, it gives the genome as common heritage a merely "symbolic" value. Hence, "the risk is great that the Declaration will serve as a flag for principles and will not prove to be an instrument used to guide practical approaches to genetic issues."[11]

Similar concerns about the meaning and effectiveness of "common heritage" rhetoric are expressed by Pilar Ossorio, who notes that it really comprises two different traditions in international law.

It is not clear what either of these would mean concretely in the case of the human genome. The first is the common heritage property doctrine, which "vests all people or all nations with equal property interests in a territory or resource," such as the sea.[12] No one person or group has a right to deprive others of the use of a common resource, for its benefits must be equitably shared. However, historically, this has not prevented commercial exploitation of the resource, such as fishing. The second is the common heritage duties doctrine, which "recognizes that certain cultural artifacts and natural features are of such importance to all humanity that all nations have a duty to help protect and conserve them."[13] The problem is that there is no one thing called the human genome that can be identified for conservation, nor is it clear that change in this resource would clearly be detrimental to those who have an interest in it. A major problem Ossorio identifies in relation to both meanings of the doctrine is the lack of a management authority that would ensure that the social and legal implications of calling the genome a "common heritage" would be met. Therefore, she asks, is there any point in using this terminology?

Recent theories of international governance support two important aspects of a reply to this question. First, even if moral ideals and norms cannot be linked in a direct and specific way to institutions and policies that realize them, they are still important in creating a public climate in which certain values and goals will be prioritized and noncompliance with widely recognized moral and social expectations will be shamed.[14] Second, the practical or institutional response to an ideal might be multifaceted, involving several institutions, policies, or agencies. In the last two decades, international institutions have become increasingly important for maintaining world order, promoting cooperation among states, and establishing bargaining mechanisms.[15] Possible avenues to effective public policy on genetic research might include a judicial approach, in which human rights codes, constitutions, and international conventions would be applied through the courts, with possible participation by public interest groups; a statutory approach, in which specific legislation is created to address new technologies; an administrative approach, relying on the gradual development of professional self-regulation; or a market-driven approach, responsive to national economic interests and to consumer choice.[16] While the shortcomings of the latter have already been indicated, it should also be noted that the judicial, statutory, and administrative approaches all present the difficulty of international scope, consistency, and compliance, as well as democratic inclusion and accountability in the process of establishing policies. The solution is clearly not going to lie in the establishment of a single world oversight body with authority to enforce decisions.

One piece of a more viable, multilateral, and multifocal solution is the emergence of a transnational society of institutions and networks, including scientists, professionals, labor representatives, human rights and environmental activists, and nongovernmental organizations, whose ability to communicate and collaborate has been greatly enhanced by the same communications media that have enabled the globalization of business and finance.[17] If globalization may be defined as "networks of interdependence at worldwide distances," it is not just economic, and does not apply only to economic institutions or to governmental and intergovernmental ones.[18] The concept of the common good in the globalization era must expand to include these new global networks of interdependence, without reversing the democratic trends of the last century and without permitting global community to sever connections with the local communities so crucial to individual identity and motivation for action.

In *Activists Beyond Borders*, Margaret E. Keck and Kathryn Sikkink trace the origins and effectiveness of "transnational advocacy networks" in three major areas: human rights, the environment, and women's rights. These networks make new resources available to actors in domestic struggles and multiply channels of access to the international system for organizations in civil society, for states, and for international organizations.[19] Key to the growth and success of such networks in organizing information, developing sophisticated political strategies, and persuading or pressuring much more powerful organizations are "the centrality of values or principled ideas" and "the belief that individuals can make a difference."[20] An example of this kind of activism in the health-care sphere is the Global Health Equity Initiative, which was started in 1995 by North American and Swedish researchers convening at the Harvard Center for Population and Development Studies. The GHEI now links over one hundred researchers from more than fifteen countries united in their commitment to address global inequities in health.[21] The concluding essay in the GHEI's publication, *Challenging Inequities in Health*, emphasizes that health equity is a problem in all societies and requires a synergistic response, in which there is two-way communication between global and local strategies, as well as cooperative leadership and advocacy at the transnational level. Changing the picture will require "a constellation of governments, ministries of health, regional organizations, nongovernmental organizations, researchers, advocacy groups, and individuals."[22]

The Catholic Common-Good Tradition

Although sensitivity to the common welfare as including the solidarity and participation of all members is certainly not limited to religious traditions, as we have seen,[23] religious voices may play an important part in enhancing social sensitivity to these values. In particular, they may function to encourage public sensibility beyond the minimal moral duties required by the principles of equality and autonomy, evoking the altruism and willingness to sacrifice self-interest that will be necessary to affirmatively remedy the situation of the powerless and vulnerable. This is notably true of the Catholic common-good tradition, given expression in the line of modern papal social encyclical, beginning with Leo XIII's *Rerum Novarum (On the Condition of Labor)* in 1891.[24] Inspired by the social problems and changes stemming from the industrial revolution, these papal letters have addressed the changing issues of the times, attempting a balance between the interests of the received hierarchical social and political order, the entrepreneurial freedoms of the new capitalist elites, and the basic human needs and rights of the working classes and the underclasses.

The moral center of this tradition is a concept of justice as including mutual rights and duties of all members and groups in society, the cooperation and interdependence of all in the common good of society, the moral and legal responsibilities of the state, and the legitimate sphere of independence of local groups and institutions (denoted by "the principle of subsidiarity"). Beyond asserting that all individuals and social groups have mutual rights and duties, the social encyclicals also identify substantive goods in which all should share, for example, food, shelter, education, employment, private property, freedom to marry and raise children, and political participation. During the 1960s and after, the concept of the common good was increasingly expanded to include a global dimension, favoring the integral development of all peoples and the ideal of a universal common good. In the 1980s and '90s, during the pontificate of John Paul II, the "preferential option for the poor" (a phrase borrowed from liberation theology) became an increasingly visible part of the common-good agenda, often imaged in gospel terms, and advanced under the ideal of "solidarity." John Paul II has been a strong critic of consumerism, materialism, and the excesses of market capitalism, as well as a defender of the human dignity of the poor. In *Evangelium vitae* (*Gospel of Life*), the pope rejects "a completely individualistic concept of freedom, which ends up by becoming the freedom of 'the strong'" (no. 19), commends greater international availability of medical resources as "the sign of a growing solidarity among peoples" (no. 26), and reminds us that "it is above all 'the poor' to whom Jesus speaks in his preaching and actions" (no. 32).[25]

A definition of the common good is offered by John Langan:

> It insists on the conditions and institutions which are necessary for human cooperation and the achievement of shared objectives as decisive normative elements in the social situation, elements which individualism is both unable to account for in theory and likely to neglect in practice.[26]

Francis McHugh claims that the basic and interdependent features of Catholic social thought are a proportionate social distribution of duties, resources, privileges, and obligations, oriented by solidarity, subsidiarity, and a preferential option for the poor.[27] Likewise, Jonathan Boswell asserts that solidarity is the peak value that guides the interrelation of core Catholic ideas such as justice and subsidiarity.[28]

The specific content or application of these ideas, values, principles, or goals must be inferred experientially and historically.[29] According to social ethicist David Hollenbach, the Catholic common-good tradition can contribute to public life and discourse by encouraging intelligent assessment of "visions of the good life" in a community of moral discourse aimed toward greater consensus about the shape of "a good society."[30] This consensus can be approached by beginning with reflection on basic human needs and goods. For example, the Aristotelian philosopher Martha Nussbaum mentions mortality, the body and its requirements, pleasure and pain, cognitive capability, practical reason, early infant development, affiliation (friendship and love), and humor as possible items on a list of common human experiences.[31] Similarly, Alan Gewirth has argued that a basic principle of equal respect must be filled out with reference to material conditions and components of well-being. These include "basic well-being" or the essential preconditions of actions such as life, physical integrity, and mental equilibrium [all of which imply basic health care]; "nonsubtractive well-being" or the general abilities and conditions necessary to maintain one's sense of purpose and take particular actions; and "additive well-being" or the conditions necessary to increase one's capacity for purpose-fulfillment and to act, such as education and the ability to acquire income. These conditions and aspects of well-being demand institutional recognition and support, though specific forms will vary culturally. McHugh argues that such an approach is represented by the Catholic and Thomistic tradition of natural law, though a term like "common social wisdom" might better capture the diachronic, communitarian, inferential, and prudential character of natural law reasoning rightly understood.[32] Nonetheless, Catholic social tradition has been equally characterized by a commitment to "truth-seeking and the pursuit of good."[33]

Bioethicist Andrew Lustig demonstrates the application of a parallel ethic in health care.[34] He stresses that social justice has an institutional and structural meaning, so that societies and governments are under a moral requirement to mediate the claims of individuals, to advance the right to medical care, to address social inequities through institutional change, and to prioritize the dignity of the most disadvantaged in society. The late Cardinal Joseph Bernardin of Chicago put health care in a social context by speaking of a "consistent ethic of life" that takes into consideration not just individual needs and decisions but the quality of life of all in a just society. In particular, he highlighted a social obligation to the most vulnerable and placed this mandate for public policy within a gospel vision.[35]

The undeniable importance of finding secular warrants for policy formation does not make religious voices irrelevant. Catholic thought can call societies to achieve a balance between rights and obligations and to curtail the excessive claims of individual rights. The universalizing tendency of the love command may remind us to reexamine the moral values already embedded in policies about access to care and can challenge the broader society to be more accountable to values that are easily submerged by a market ethos, though in principle recognizable to all.

One endemic shortcoming of the past Catholic common-good tradition is that, while it has increasingly turned attention to the plight of the poor, it has usually still conceived of social change as emanating from the top down and has directed its calls for change primarily at those enjoying power. Empowerment of the poor and their inclusion as decision makers is a theme of Marxist social critique and of liberation theology that is beginning to make inroads into Catholic thought about the common good and health care. The Brazilian theologian and bioethicist Márcio Fabri dos Anjos sees it as fundamental that the poor gain consciousness of their fundamental rights and responsibilities and come into the ethical arena as "co-workers, not needing to remain in their customary condition of recipients of compassion."[36] As Fabri notes, however, this participatory stance is not always readily achieved, precisely because of the absence of interpersonal and institutional "solidarity" between the beneficiaries and the victims of long-lasting colonial relationships. One symptom is the absence of laws and regulations governing research in many developing nations, allowing the scientist to arrive and operate "with no rules other than those he makes himself."[37]

Another challenge to the traditional Catholic concept of common good is that it assumes a concept of world order in which concentric circles of society and law move from the most local to the most global, with final authority inhering in a centered, comprehensive, and impartial system focused, for example, through a worldwide public authority such as the United Nations.[38] The hope for such an authority is clearly obsolete. Can the traditional conceptual framework cope with the realities of globalization and the new decentered forms of international governance? Though developed within the old unitary and hierarchical framework, the "principle of subsidiarity" offers a possible means of renegotiating the Catholic common-good idea. Originally developed to defend the right of smaller bodies or communities from state interference, this principle came to be used by John XXIII to defend the obligation of the state or the world authority to intervene against lower-level injustices when necessary.[39] In the context of the search for world peace, John states the principle in a way that moves toward a view of the global social order in which many types of social organization can operate in networks of collaboration. What needs further revision today is the assumption that these are hierarchically related in an overarching common order.

> The worldwide public authority is not intended to limit the sphere of action of the public authority of the individual state, much less to take its place. On the contrary, its purpose is to create, on a world basis, an environment in which the public authorities of each state, its citizens, and intermediate associations, can carry out their tasks, fulfill their duties and exercise their rights with greater security.[40]

The new call for power sharing in solidarity can be linked to a more participatory and egalitarian notion of subsidiarity.[41] As the following section will illustrate, statements of the pope, Vatican representatives, and Catholic activists on international patenting law show that at least at the practical level, the more diverse and diffuse model of international governance appearing in political science literature is also reflected in recent Catholic common-good argumentation.

Patents and the Common Good

International patent law has gradually expanded to include the right to patent discoveries of genes and gene segments, even before any new product or procedure based on knowledge about the gene has been invented. The 1980 Chakrabarty case in the United States allowed a life form, an oil-eating micro-organism, to be patented because it was a human invention, encouraging a similar trend in Europe and Japan to allow the patenting of life forms. Patent laws tend to protect those with research resources and ensure their ability to make more income rather than ensuring the availability of their research fruits to the poor, even if the poor have participated in the development of knowledge by serving as research subjects or by otherwise providing expertise and knowledge that contribute to the progress of research. Corporations are increasingly gaining control, through investment-funded research and patents, over the new technologies of genetic modification, whether in the food supply, in humans, or in pharmaceuticals. Márcio Fabri mentions several cases, in one of which a U.S. corporation claimed to have found a cure for asthma in an African coastal tribe, then sold DNA samples from tribe members to a German pharmaceutical company for 70 million U.S. dollars without any remuneration whatsoever to the DNA donors. "This little example shows that genetics has become a field of economic and political endeavor that national and international policies cannot ignore."[42]

A report on biopatenting and food security put together by CIDSE (International Cooperation for Development and Solidarity), a network of fifteen nongovernmental development organizations from Europe and North America, represents an effort to bring concerns about the common good and solidarity into the international forum of public policy about genetic science.[43] The purpose of the report is to help bring trade rules into convergence with international development targets as set by a series of UN conferences in the 1990s, including the 1996 World Food Summit. It recommends that all life forms be excluded from patenting, including plant varieties. A key concern of CIDSE is that "Transnational corporations (TNCs), that is, big agrochemical or pharmaceutical corporations, have become the driving force behind genetically modified food, the global spread of industrialized agriculture and the privatisation of knowledge, bringing intellectual property rights under the WTO and its legally binding dispute settlement procedures."[44] As a concomitant, the poor cannot exercise full agency in helping to determine issues that affect their lives. "They are denied this by the poverty, hunger, disease and lack of education which afflict so many in the South."[45] Here surfaced in the context of genetically modified food, these concerns will apply with even greater force to the genetic modification of human beings.

CIDSE refers to the Word Trade Organization's Trade Related Aspects of Intellectual Property Rights Agreement (TRIPS) as setting international policy on patents but resists the total legitimacy of this agreement because of its undemocratic character, its effects on the poor, and its permission of patenting on life forms that should be considered the common heritage of humankind. However, it urges WTO member nations to make exceptions to the agreement on the basis of TRIPS article 27, which provides that some things may be excluded from patentability in order to protect public order or morality. CIDSE asserts that "no human being has an absolute right of ownership," and places patent rights within a larger framework of responsibility and the common good.[46] It refers to a speech of John Paul II, made shortly before 1999 IMF and World Bank meetings, in which he calls for debt relief and investment in education and health care. The pope invokes the church's teaching that "there is a 'social mortgage' on all private property, a concept which today must also be applied to 'intellectual property' and to 'knowledge.' The law of profit alone cannot be applied to that which is essential for the fight against hunger, disease and poverty."[47]

In June 2001, during a meeting of the council for TRIPS, Archbishop Diarmuid Martin, Vatican observer to the WTO, delivered a text representing the Vatican's position. He took as his point of departure the limited availability, especially in Africa, of patented AIDS drugs, and insisted that, although patent rights are important to provide just compensation and promote research, they must be part of a broader framework of analysis. Pricing drugs out of the reach of the poor sacrifices the common good to financial gain.

> The unity of humankind and the universality of human rights (among which the right to health) require that all the economic and political actors involved (international organizations, governments, private foundations, corporations and NGOs) work together, pooling their differentiated responsibility for resolving a global crisis, leaving aside narrow individual or sectorial interest.[48]

The conflict between financial interests and the needs of those lacking access to medicines must be based on "a broad-based commitment of solidarity" to development and to the common good. The problem and its solution go beyond the responsibility of the council for TRIPS, and becomes "the common responsibility of many other international organizations as well as national governments, and in an appropriate manner also of the private sector."[49]

A dispute over the availability of AIDS drugs in South Africa began a series of developments that in the year 2001 resulted in the beginning of a multifaceted challenge to and revision of the patent regime that excludes poor populations from access to life-saving treatments. After the South African government enacted a law permitting the importation of cheap generics, the South African representative of thirty-nine international pharmaceutical companies brought suit against the government in a national court. After enormous negative publicity and offers from competitors to provide the drugs at low cost, the suit was dropped, and offers of inexpensive or even free drugs increased, even from some of original complainants. In November, World Trade Organization members agreed to an interpretation of its 1994 TRIPS agreement that permits nations to determine their own standards for determining when a national emergency exists and for permitting the manufacture or even importation of generic versions of patented drugs. This development still leaves unresolved the question of how very poor countries are to pay for local drug manufacture and distribution, but it is a move in the direction of local decision making and sharing of power.

Conversion to Compassion, Altruism, and Solidarity

A key plank in the platform of CIDSE is the preferential option for the poor. "The Gospels require us never to overlook the poorest of the poor and most vulnerable members of society, telling us again and again that they are our sisters and brothers, and more, to ask not only what impact any particular measure will have on them but to discern what they have to contribute in insight, learning and inventiveness."[50] Not limiting this mandate to Catholics or Christians, the CIDSE report mentions Gandhi's motto that the ethics of any action should be decided keeping before one the face of the poorest of the poor. The American scientist Bill Joy appeals to essentially the same moral sensibility to help avoid disasters brought about by uncritical genetic "progress," calling for "fraternity" founded on altruism, for "fraternity alone associates individual happiness with the happiness of others, affording the promise of self-sustainment."[51] Two Canadian legal scholars, Bartha Maria Knoppers and Claude M. Laberge, likewise recommend that genetic policy be governed by principles of reciprocity, mutuality, solidarity, and universality.[52] Bioethicist Dan Brock goes so far as to assert that "it is a commonplace in general theories of justice that special concern is required for the worse-off members of society, and that the justice of a society can be judged by how it treats its least well-off members."[53]

While solidarity and altruism are recognizable ideals to many, it is not always easy to give them functional priority when corporate investments or the privileges of social groups are at stake. Religious traditions can keep these priorities in public view, working to shape or reform consciousness of the fact that the common good requires participation by all members. One important way in which they do this is through the images and stories that concretize the needs of the poor and give them the "face" imagined by Gandhi. Even in an age of globalization, what most motivates us to action is what is at stake locally, in the communities and networks of "everyday life experience."[54] If formulations of justice and equity are abstract, they lack power to compel assent, commitment, and action. Philosophical bioethicist Daniel Callahan considers "the decline of religious contributions a misfortune, leading to a paucity of concepts, a thin imagination, and the ignorance of traditions, practices and forms of moral analysis of great value."[55] The narratives, symbols, saints, and parables of religious traditions can broaden the moral horizon of public discourse by inducing hearers (even unbelievers) to place themselves imaginatively at the side of the poor, giving life to what may otherwise remain a bloodless and unmotivated intellectual recognition of equality, easily dislodged by perceived threats to the immediate condition of one's "life world." Unless political and social agents can envision those who are socially "other" as in some way part of a shared life world, the political will to change social practices or to establish new ones on their behalf will ultimately be lacking.

Summary

Individualism, autonomy, and market economics lead to a biotechnology environment in which human genetics may become just another axis of social inequality, yet one with immense potential to write social inequities into the human body, perhaps perpetuating them for generations. Although formal recognition may be granted to justice as equal respect, economic incentives to override this principle at the practical level are huge. The individualism of market entrepreneurs and consumers needs to be corrected by a greater personal and public sense of the common good, which establishes inclusive practices of decision making and also identifies essential substantive goods like food and basic health care. The Catholic common-good tradition (and other religious traditions) can bring to the public forum a viable and persuasive concept of the common good as entailing full social solidarity, participation, and empowerment, as well as the right to basic material and social goods. Moreover, it can embody a concept of an inclusive common good in specific narratives, images, and practices that capture the imagination and help create the political will to change social structures and institutions.

The current situation of a decentralized and mobile process of world government engaging many intersecting international institutions and networks provides an unprecedented challenge to the traditional hierarchical Catholic concept of the common good. Yet there are resources in the tradition for meeting this challenge, particularly a renewed and extended principle of subsidiarity. The vitality and effectiveness of principled activism for justice worldwide also lends credence and persuasiveness to the characteristically Catholic confidence that action for positive social transformation can in fact bring about change and create a better society. Although both social inequities and social activism exist in many sectors of society, the emergence of the new science of genetics in coordination with economic globalization creates an unparalleled opportunity to reexamine old structures and interactively create social policies that better serve the common good of all societies, groups, and persons.

Notes

1. Portions of this essay appeared in an earlier version as "Genetics, Individualism and the Common Good," in *Interdisziplinare Ethik: Grundlagen, Methoden, Bereiche,* ed. Adrian Holderegger and Jean-Pierre Wils (Freiburg: Universitätsverlag and Herder, 2001), 378–92.

2. See Joseph S. Nye, Jr. and John D. Donahue, eds., *Governance in a Globalizing World* (Cambridge, Mass. and Washington, D.C.: Visions of

Governance for the 21st Century and Brookings Institution Press, 2000); and Timothy Evans, Margaret Whitehead, Finn Diderichsen, Abbas Bhuiya, and Meg Wirth, *Challenging Inequities in Health: From Ethics to Action* (Oxford: Oxford University Press, 2001).

3. Timothy Caulfield, "Regulating the Commercialization of Human Genetics: Can We Address the Big Concerns?" in *Genetic Information,* ed. Alison K. Thompson and Ruth F. Chadwick (New York: Kluwer Academic, 1999): 150.

4. Ibid., 150–53.

5. UNESCO: *Universal Declaration on the Human Genome and Human Rights*, UNESCO Document 27 V/45, adopted by the Thirty-First General Assembly of UNESCO, Paris, November 11, 1997, in *Journal of Medicine and Philosophy* 23, no. 3 (1998): 334. Article 12 similarly states, "the applications of research, including those in biology, genetics and medicine, concerning the human genome, shall seek to offer relief from suffering and improve the health of individuals and humankind as a whole."

6. Ibid., 335, 338.

7. Ibid., 336.

8. Christian Byk, "A Map to a New Treasure Island: The Human Genome and the Concept of a Common Heritage," *Journal of Medicine and Philosophy* 23, no. 3 (1998): 236.

9. See Susanne Boshammer, Matthias Kayss, Christa Runtenber, and Johann S. Ach, "Discussing HUGO: The German Debate on the Ethical Implications of the Human Genome Project," *Journal of Medicine and Philosophy* 23, no. 3 (1998): 327.

10. Byk, "A Map to a New Treasure Island," 241–42.

11. Ibid., 245.

12. Pilar Ossorio, "Common Heritage Arguments Against Patenting Human DNA," in *Perspectives on Genetic Patenting: Religion, Science, and Industry in Dialogue,* ed. Audrey R. Chapman (Washington, D.C.: American Association for the Advancement of Science, 1999): 92.

13. Ibid., 96.

14. Robert F. Drinan, S.J., *The Mobilization of Shame: A World View of Human Rights* (New Haven and London: Yale University Press, 2001).

15. Robert O. Keohane, "International Institutions: Can Interdependence Work?" in *Foreign Policy* (1998): 82–83.

16. Bartha Maria Knoppers, Marie Hirtle, and Kathleen Cranley Glass, "Commercialization of Genetic Research and Public Policy," *Science* 289 (1999): 2277–78.

17. Keohane, "International Institutions," 93.

18. Joseph S. Nye, Jr., "Globalization's Democratic Deficit: How to Make International Institutions More Accountable," *Foreign Affairs* 80 (2001): 2.

19. Margaret E. Keck and Kathryn Sikkink, *Activists Beyond Borders:*

Advocacy Networks in International Politics (Ithaca/London: Cornell University Press, 1998): 1.

20. Ibid., 2.

21. Evans et al., "Preface," *Challenging Inequities*, ix.

22. Margaret Whitehead, Göran Dahlgren, and Lucy Gilson, "Developing the Policy Response to Inequities in Health: A Global Perspective," in *Challenging Inequities*, 322.

23. See also the "liberal communitarian vision" of Ezekiel J. Emanuel in *The Ends of Human Life: Medical Ethics in a Liberal Polity* (Cambridge, Mass./London: Harvard University Press, 1991).

24. For a general presentation and discussion, see John A. Coleman, S.J., ed., *One Hundred Years of Catholic Social Thought: Celebration and Challenge* (Maryknoll, N.Y.: Orbis Books, 1991).

25. John Paul II, *The Gospel of Life* (Boston: Pauline Books, 1995).

26. John Langan, S.J., "Common Good," *Westminster Dictionary of Christian Ethics*, ed. James F. Childress and John Macquarrie (Philadelphia: Westminster Press, 1986): 102.

27. Francis P. McHugh, "Muddle or Middle Level? A Place for Natural Law in Catholic Social Thought," in *Catholic Social Thought: Twilight or Renaissance?* ed. J. S. Boswell, F. P. McHugh, and J. Verstaetern (Leuven: Leuven University Press/Uitgeverij Peeters, 2000): 51.

28. Jonathan S. Boswell, "Solidarity, Justice and Power-Sharing: Patterns and Policies," in *Catholic Social Thought*, 104–5.

29. McHugh, "Muddle or Middle Level?" 51–52.

30. David Hollenbach, S.J., "Afterword: A Community of Freedom," in *Catholicism and Liberalism: Contributions to American Public Philosophy*, ed. Bruce Douglass and David Hollenbach (Cambridge/New York: Cambridge University Press, 1994): 324, 334. Hollenbach also delivered a presentation on "Globalization, the Market, and the Common Good" at the American Academy of Religion, 1998, in which he presented the preferential option for the poor as a part of the U.S. bishops' approach to economic justice.

31. Martha C. Nussbaum, "Non-Relative Virtues: An Aristotelian Approach," in *The Quality of Life,* ed. Martha C. Nussbaum and Amartya Sen (Oxford: Clarendon, 1993): 263–65.

32. McHugh, "Muddle or Middle Level?" 53, 55–56.

33. Ibid., 44.

34. B. Andrew Lustig, "The Common Good in a Secular Society: The Relevance of a Roman Catholic Notion to the Healthcare Allocation Debate," *Journal of Medicine and Philosophy* 18 (1993): 569–87.

35. Joseph Cardinal Bernardin, *A Moral Vision for America*, ed. John P. Langan, S.J. (Washington, D.C.: Georgetown University Press, 1998).

36. Márcio Fabri dos Anjos, "Power, Ethics, and the Poor in Human

Genetics Research," in *The Ethics of Genetic Engineering*, ed. Maureen Junker-Kenny and Lisa Sowle Cahill (London/Maryknoll, N.Y.: SCM and Orbis Books, 1998), 82.

37. Ibid., 78.

38. John XXIII, *Pacem in terris* (*Peace on Earth*), in David J. O'Brien and Thomas A. Shannon, eds., *Catholic Social Thought: The Documentary Heritage* (Maryknoll, N.Y.: Orbis Books, 1998), no. 138. A contemporary version of this kind of expectation is represented by the call of one Catholic ethicist for a "transnational monetary and financial order" that includes a transnational legislator, a transnational executive, and transnational supervising bodies (Jef Van Gerwen, S.J., "Global Markets and Global Justice? Catholic Social Teaching and Finance Ethics," in *Catholic Social Thought*, 218–19).

39. See Pius XI, *Quadragesimo anno*, no. 79; John XXIII, *Mater et magistra*, no. 53; and John XXIII, *Pacem in terris*, nos. 140–41, all in O'Brien and Shannon, *Catholic Social Thought,* 60, 92, and 153–4, respectively.

40. John XXIII, *Pacem in terris*, no. 141 (p. 154).

41. Boswell, "Solidarity, Justice and Power Sharing," 107.

42. Fabri, ""Power, Ethics, and the Poor," 74.

43. CIDSE, *Biopatenting and the Threat to Food Security: A Christian and Development Perspective*, ed. Bob van Dillen and Maura Leen. Available from CIDSE General Secretariat, Rue Stevin 16, 1000 Brussels, or on the internet at http://www.cidse.be/pubs/tp1ppcon.htm.

44. Ibid., 1.

45. Ibid., 2.

46. Ibid., 4.

47. Pope John Paul II, "Jubilee Year Calls for Urgent Debt Relief," *L'Osservatore Romano*, September 23, 1999, English ed. (September 29, 1999) 2; or www.ewtn.com/library/PAPALDOC/JP2DEBT.HTM.

48. Archbishop Diarmuid Martin, "Statement on Trade Related Aspects of Intellectual Property Rights," *L'Osservatore Romano*, English ed., July 11, 2001, 9; or www.ewt.com/library/CURIA/statmdp.htm.

49. Ibid., 9.

50. Ibid., 4.

51. Bill Joy, "Why the Future Doesn't Need Us," *Wired* 8, no. 4 (April 2000). Available on the internet at http://www.wired.com/wired/archive/8.04/joy_pr.html. Joy is cofounder and chief scientist of Sun Microsystems, Inc., and cochair of the U.S. President's Information Technology Advisory Committee.

52. Bartha Maria Knoppers and Claude M. Laberge, "Ethical Guideposts for Allelic Variation Databases," *Human Mutation* 15 (2000): 30–35.

53. "Broadening the Bioethics Agenda," *Kennedy Institute of Ethics Journal* 10, no. 1 (March 2000): 26.

54. Arthur Kleinman, "Moral Experience and Ethical Reflection: Can Ethnography Reconcile Them? A Quandary for the 'The New Bioethics,'" *Daedalus* 128, no. 4 (Fall 1999): 70.

55. Daniel Callahan, "The Social Sciences and the Task of Bioethics," *Daedalus* 128, no. 4 (Fall 1999): 280.

7

Power and Vulnerability: A Contribution of Developing Countries to the Ethical Debate on Genetics

MÁRCIO FABRI DOS ANJOS

The social and political history of genomics in developing countries may serve as a mirror in which to view the major ethical challenges for the world in this field. In such a context there are islands of high technology with research and services and great economic interests floating on oceans of basic lack and poverty. Power and fragility are neighbors. Human genomics offers a good test of humanity's ability today to work out ethical questions great and small.

We offer here an approach to the Latin American context, looking for constructive contributions to ethical reflection on this subject. Our emphasis will be on the social dimensions that present major challenges to the ethical evaluation of human genomics, recognizing that a complete analysis would require attention to many other aspects. We will collect Latin American elements to enrich the bases of a Christian contribution, developing human genomics in a participative way with vulnerable and poor people. We will conclude with some comments on important social spaces for the construction of ethics as reflection and as practice in the context of inequalities.

Some Introductory Remarks

To introduce human genomics from the Latin American context, we will draw first a generic picture of progress in two corresponding areas: genetics research and its results. Both converge in services, in the form of diagnoses, preventions, improvements, and therapies. Obviously, many services start from germ-line and reproductive medicine. Among the beneficiaries must be counted not only individuals but also social groups, especially in the case of epidemiologies, and the whole society, in the present and in the future, particularly if we consider genetic alterations that are transmitted to new generations. The enterprise of development consists in two types of initiatives, public and private, with their corresponding interests.

Ethical evaluation is required in each of these areas. In research, the major concerns of ethics are the subjects and objects of research, the objectives and the researchers' interests, and the methodology of the procedures. When results are presented as *services*, ethical concerns are turned mainly to effects and consequences and to the quality of social distribution of benefits and risks. Evaluation, proposals, and predictions are important tasks of ethics in this area where the future of humanity is at stake. Finally, it is necessary to remember that the elaboration of ethics depends, in its foundation and criteria, on an atmosphere favorable to ethical reflection, that is, on social conditions that encourage and sustain ethical discourse. Autonomy, for example, is today a widely appreciated condition and norm for the construction of ethics.

In the Latin American context, all these topics fall within a field wherein great scientific conquests intersect with equally great social ambiguities. How far does autonomy exist as a real condition of the ethical reflection of most people? How are people informed about genomics, and does information become a way to dialogue, a bridge to participation in debates, a significant step to education about new technologies? In the Latin American context, the vulnerability of many people and the lack of the social conditions necessary to a participatory construction of ethics in genomics create big challenges that must be taken into account. Efforts being made to reverse this situation, however, must be also emphasized.

A Context for the Production of Genomics and for Ethical Reflection

At first sight, one might think that the limited resources of developing countries would almost prevent real scientific participation in technological research. Such countries would be mainly admirers of development in genetics, spectators of the ethical discussions, and merely occasional consumers of technical results and ethical conclusions. But the growing participation of Latin America in the development of genomics is surprising. Among the world's countries, Brazil ranks second in the contribution of information in the screening of the human genome. On chromosome 22 alone, Brazil's contribution represented a third of the total data found. The country developed its own methodology. "More than 95% of the fragments of identified genes in tumors of head and neck were sequenced by the Hospital of Cancer in Sao Paulo."[1] The country is among the best in the world in the development of applied genetics for tropical agriculture. Brazil has had excellent results in the sequencing of noxious bacteria, in the development of defensive agricultural biology, and in the sequence and improvement of products such as sugar cane.[2] This country counts more than 11,800 research centers, of which more than 30 percent are dedicated to research directly linked to health.[3] It is not necessary to expand further on the evidence but only to emphasize that developing countries are not necessarily outsiders to significant research and to ownership of new technologies in genetics.

It is also true, however, that this participation in research occurs on a continent that struggles with great lack of resources. Who invests in the researches and in the production of services in genetics in such a situation, and for what purposes? Much investment has its source in state resources, usually channeled to research centers linked to universities. But private initiatives from within and outside the country have a growing presence. Recent data on protocols submitted for ethical evaluation to the national commission indicate an increase, in the year 2000, of almost 100 percent in research projects in Brazil supported with foreign cooperation, and 88 percent of these involve new drugs.[4] This reveals market interests and the familiar strength of the pharmaceutical industries.

An Atmosphere for Ethics in Human Genomics

The general lack of resources, added to the disparity of economic forces, creates an atmosphere of enormous inequalities and social injustice on the Latin American continent. To understand the reach of this fact in the elaboration of ethics in human genomics, it is necessary to analyze the social construction of inequality. In other studies, we explored the political-economical aspects of this inequality, and we showed the vulnerability of individuals, of social groups, and even of whole nations under such conditions.[5] Here, we would like to highlight some cultural bases of the Latin American context as providing the atmosphere for the building of ethics in human genomics.

The cultural bases of a people are an indispensable element of ethical reflection. Genomics in particular provokes an enormous revolution in the main meanings of life and in their constructed social forms. We know that cultures are not static but are in continuous transformation. They change through interaction among groups and in contact with nontraditional values. These dynamics generate a social ethos that grounds ethical elaboration.[6]

In Latin America, as elsewhere, social inequalities are built through a long historical process that establishes the social ethos. Three factors are particularly connected in the composition of this ethos. These are (1) the great changes in the forms of production of life, passing through industrialization and culminating in our *technological* times; (2) the network of human and environmental *relationships* in which this production of life occurs; and (3) the elaboration of *meanings* given to all components (persons, animals, and things) in this process of change.

Since the time of colonization, the largest part of the population was systematically dominated. Indigenous people, slaves, and popular segments were marginalized from partaking in decisions about public life, as well as from important ways to it, like scholarship and education. Disadvantaged in the social network of relationships, a significant part of the population has no possibility of participating appropriately in technological development and in the construction of meanings necessary to guide human life in this new context. Sociocultural analyses speak today of an increasing process of *exclusion*, an appropriate concept to understand the condition of the poor in the network of social relationships, understood in light of the conjugation of the three factors above. In the Latin American context today, it would be difficult to understand social injustice without awareness of a cultural inheritance marked by exploration, domination, and exclusion. The injustices and inequalities no longer evoke repugnance or fear because they have become historical and "traditional." Is the planet as a whole experiencing the same situation?

Thus we can say that the development of human genomics as part of technological progress integrates a wider process of social asymmetries. Ethical procedures in genomics are today adversely affected by this established ethos. The tendency for this type of society to prevail puts profit above people, and ethical proceedings are seen as an obstacle to the facility of research as well as for development in general. A recent statistic shows, for instance, that 79.1 percent of 139 scientific magazines in Brazil do not request, in the instructions for their authors, ethical references on research procedures.[7] Neither is it by chance that, until now, most of the Latin American countries have not had commissions for research ethics. Nevertheless, there are also many successful initiatives in the area of genetic ethics that will be mentioned below.

To be exact, we should say poor people and poor groups are not excluded absolutely from the development of human genetics. They participate in several ways, for example, as subjects of research and as receivers of information. Directly or indirectly, even if in the long term, they will participate in the results and risks for humanity arising in this field. But the difference is in the vulnerable participation they have, a vulnerability that can be exploited when convenient. A consultation developed by the WHO on genomics on the American continent accentuates the need to place the ethical principles "in the specific social, cultural and political contexts of American countries. This has to start with a frank diagnosis: eugenics has occurred in Latin America; the discrimination of the handicapped is widespread; there is much left to do in the field of reproductive rights."[8] For ethics, the exploitation of vulnerability can extend beyond direct violation of human dignity, causing an exclusion of important partners to ethical deliberation and exposing the fact that the theoretical construction of ethical criteria occurs under the interests of those who have technological power.

Opting for the Poor in Human Genomics? Some Theological Contributions

Ethics and genomics can benefit from a contribution of Latin American theology in its concern for the poor. Asking the question of the place of the poor in genomics can provide criteria for ethical evaluation and proposals. Since the connection between faith and science is much appreciated in liberation theology,[9] we must include in the ethical sphere today scientific discourse and the prominence of accomplishments from the empirical sciences. An important question in this framework is how to deal in a humanitarian way with such powerful tools, especially since a great part of humanity is in need.

In order to develop this theme, we will note, first, some theoretical aspects of the concept of *power* in the reading of liberation theology. This will provide a preliminary understanding of an "option for the poor in genomics." Second, we will comment on some concrete initiatives in the Latin American social context, where the question of participation of the people in the ethical assessment of genomics is at stake.

The New Face of the Poor in Technological Times

A dissertation I directed calls attention to the social and cultural revolution that is happening in the life of poor populations amid present technological developments.[10] The study analyzes the impact of great technological changes in the cultural and religious *Weltanschauung* of the Mayan people in Guatemala. How can this rural and poor people be introduced fairly to the new network of relationships brought about in these biotechnological times? At the same time, we can wonder whether it is urgent for this people to rework the symbols, myths, and meanings[11] that compose their *Weltanschauung* to meet the crisis of technological knowledge.

There are, therefore, new challenges to understanding the poor in our day. Culturally, they are touched in the shrine of their symbols. They are in need not only in an economical sense but also culturally. They need help in the reconstruction of their own identities—the symbols and meanings that sustain their subjectivities and that compose the new traces of their cultures. How difficult it is to deal with autonomy in such a context?

Theology in Latin America, even if conscious of the growth of cruelty in terms of poverty, tries to analyze and to understand the wider and complex context in which the poor live.[12] It does not help, nor is it evangelical, to identify technological and scientific development as evil (a "satanization" of the new social context as a whole). More in opposition to the evangelical experience than technology as such is the concentration of *power*—in old and new ways—that excludes the poor. There is a dynamic reproduction of poverty. Inside this picture, it seems extremely important that theological reflection accompany scientific and technological progress, creating new forms of understanding of human beings living together as partners on the planet. The poor must not be forgotten, nor can they be reduced to a mere accident of history.

Regarding the relation of the poor to the development of human genetics, theology will situate scientific progress in the group of human and environmental relationships built by the great social options. L. Boff says that "in the original option for the poor should enter, first, the great poor, the earth and humanity; on this base, if safe, we could put the question about the future of the poor and of the condemned people of the earth. All of us need to be liberated . . . from a paradigm of civilization, today globalized, that can destroy us collectively."[13] In another text, always with his vigorous language, Boff designates some lines of this threatening process:

> the competition in the economy and in the market, made a supreme principle, strangles the cooperation necessary to the common possibility of living and to development. Neoliberal thought, imposed as unique, destroys the cultural and spiritual diversity of the people. There is an imposition of a single way of production, with the use of a single type of technology and of a single administrative model, maximizing the profits, shortening the time and minimizing the investments. It devastates the ecosystems and puts under risk the living system of Gaia, the Earth.[14]

The key framework for ethical questions about the development of genomics is, therefore, questions about social life and environmental systems.

Power: A Meeting Point of the Poor and Genomics

Power is a key concept in the development of genomics. It can be taken in a technological sense to express the energy of progress; in the economical sense, to refer to the force of investment in research and in the negotiation of their results; in the political sense, to refer to the capacity to coordinate other powers; in the cultural sense, to denote the fascination and enchantment exercised against the imaginative horizon of humanity. Power has multiple faces and many forms. Knowledge is a privileged form of power because "it generates power of highest quality."[15] The development of human genomics depends on the exercise of several forms of power. And at the same time, this development multiplies power in our hands.

As a contribution to an ethical approach to genomics, we would like to introduce here the concept of *power* reflected in liberation theology. Because the concept of power is not expressed in an abstract way but applied to relationships, different terms are required to express the diverse moral connotations of its exercise. In this sense, we can see that power is really a meeting point of the poor and the development of human genetics.

Inspired by biblical narratives, liberation theology emphasizes the understanding of power in the context of human relationships, especially between the strong and the weak. A concept of power in an absolute form is attributed to God, and here begins the lesson for humans. The absolute power of God is as creator and communicator of potentialities. It is a power that aims for empowerment. There are three fundamental aspects of power understood in this sense, with a fourth aspect opposed to them:

1. *Power-empowerment:* This concept is expressed by the Greek term *exousia* (ἐξουσία),[16] which can be understood as a communicated and communicative power. "According to the synoptic gospels, Jesus conducted his ministry on the basis of the divine power imparted to him through the Holy Spirit. . . . At the same time, Jesus bestowed his power and authority on his followers."[17] In modern terms, *exousia* suggests the capacity to potentiate (capacity to empowerment)—power that is exercised in favor of others, enlarging their possibilities to be and to act. This concept becomes clearer if we examine its corresponding Latin term, *augere*, which is translated as *authority*. In its etymology, *augere* suggests a capacity to enlarge, to facilitate. In the religious language of the Gospel, the expression translated as "power," attributed to Jesus, is taken thoroughly in this sense, and corresponds very often in the original text to the term *exousia*.[18]

This *power* is communicated to the disciples in order that they can empower all human beings. Thus, "the crowd was filled with fear and glorified God who gave such a power (*exousia*) to humans" (Matthew 9:8).

2. *Power-dynamics:* A second concept of *power* complements the concept of *exousia*; it emphasizes this concept as a moral force. It shows the opposition to destructive forces and enhances its legitimacy in the construction of the common good. The term *dynamis* in the Gospel texts in this sense is easily associated with *exousia,* as in the following text: "He gave them [to the disciples] power and authority[19] over all the demons and the gift of curing illnesses" (Luke 9:1).[20]

3. *Power-sign:* If power is taken in the sense of doing good to others, it is not difficult to reach the concept of power as *sign*. It means basically that power is exercised in a progressive way, enlarging possibilities and generating life and its expressions. A *sign* says that the *exousia* is starting to act. It is at the same time demonstration and promise. It announces that something more grandiose is happening. The *power-sign*, in the religious language of the Gospel (*semeion*—σημεῖον), is associated with the concept of "miracle."[21] But in this sense, far from being a show, the miracle appears like an indication of *power-exousia*, a power that starts a process of empowerment. In the Gospel, miracle and sign are frequently equivalent terms; and cures of diseases are used as symbols of spiritual transformation.[22]

4. *Power-domination:* A fourth concept of power implies domination by force. This sense of power is opposed to the three previous concepts. It recalls the use of physical force or moral impositions. The Greek term *kratos* (κρατός) is normally applied in the Gospel texts to express this concept. However, it is not used to describe an action of Jesus; and its use is less frequent in the New Testament.[23] The use of *kratos* is opposed to the way of acting assumed by Jesus and strongly recommended by him to his disciples.[24]

The different concepts of power that can be denoted by the general term can ground a global proposal for ethical reflection about the development of human genetics. Obviously, this does not make superfluous a further definition of the meaning of constructive empowerment or of the benefits for a people or for humanity that genetics can offer. But it indicates the foundations of solidarity and commitment that are necessary to oppose the race for dominant and exploiting power in this area. In this perspective, the poor can taste the development of genetics as a sign of grandiose things happening for the good of all.

Poverty, Perfection, and Genomics

Completing our theological contribution on this topic, we find another interesting concept to analyze: perfection. It is not necessary to enter into a thorough philosophical discussion but only to draw some connections to genetics and to power. In a certain sense, perfection is a general goal that impels all the efforts in genomics. Through the results of genomic research, one seeks a development of perfection of the human being and of humanity's quality of living. A search for technological power can be understood as a search for perfection in the face of human limits and vulnerabilities. Enhancement and selective applications are, in this sense, well-known controversial types of the search for perfection.[25]

What does Christian theology have to say about perfection in this context? We do not intend to consider here specific questions like enhancement and selective applications. It seems also unnecessary to recall all the criticism and proposals that Christian theology has developed during its history about this concept.[26] Our contribution will consist in reflections on the biblical roots of this concept and some indications of the relevance of human perfection to genetics.

Briefly, we note that the concept of "perfect" in the Old Testament (*tamin*) is not applied directly to God but only to divine actions and works.[27] This suggests that human perfection likewise is found through a process of action before God. If applied to men and women, the term means a religious integrity in a way of life directed toward God. In this sense, the term will be used as a "synonym of perfect obeisance of the Law."[28] It means a *fully developed* person in a moral sense.[29] We could say it takes the way as the destination, because God in his fidelity will conduct those who are living according to the divine commandments to a final term, that is, to *perfection.*

At this point, to understand the meaning Jesus gave to the concept of perfection, it is necessary to recall first his criticism of the legalistic conception of perfection as a *perfect* obedience to the legal prescriptions collected from the interpretation of the Law and from the "code of purity."[30] According to that legalistic meaning, only those who follow all the prescriptions are perfect. This becomes a reason to discriminate against poor people, who have difficulty obeying that code. Disease, illness, and other problems of health and economic life are viewed mostly as a punishment for sins and for human imperfections.

The criticism of Jesus against that conception is simple but incisive. He puts all kind of needs, human imperfections, and even sin as a challenge for communicative power and not as an obstacle. For example, to answer why a man is blind, he does not point to the "cause" but emphasizes the chance to manifest in him the "action of God."[31] We would say he proposes perfection as a process of *exousia*, as a constructive way to oneself and to our neighbors.

In this sense, the well-known expression "be perfect as the Father is perfect" appears in the Gospel of Matthew.[32] It is formulated as a criticism to a legalistic comprehension of the Law and, at the same time, as a proposal of perfection as love. Note here that the face of God is presented as Father, which is analogous to saying that love and mercy are the essence of perfection. More directly, according to Matthew, Jesus said to the rich young man: "If you want to be perfect, sell all you have and give it to poor people. . . ."[33] Because it was not necessary in the context of the Luke's Gospel to evoke criticism of the legalistic conception of perfection, Jesus says incisively: "Be merciful as the Father is merciful."[34] A personal perfection is proposed in terms of relationships and not in an isolated relation of the individual to God. Without fear of personal losses, perfection should include sharing one's life, not rejecting the imperfect, but welcoming him or her with love, for his or her transformation.[35]

At first sight it seems that these data have nothing to do with ethics and genomics. Some important questions are present, however. One of them concerns the concept of perfection that presides over the contemporary development of genetics. Our current cultural moment pays tribute to some important factors that characterize our type of progress. For instance, with the accentuation of individualism and with the development of modern machines, we mold a new paradigm of human perfection as well as of nature.[36] Human perfection is pragmatically associated with efficiency; and competition becomes the arena in which to show it. If not profitable to the interests of the market, needs and lacks are detestable; they should be eliminated, because they mean a defeat.[37] Humanity's perfection as the development of a capacity to love does not mean rejection but offers welcome and help. Enhancement could be critically considered as a consequence of love and not as a prior condition for it.

We have here the proposal of a spirit to guide the development of genetics. In this spirit, the basic questions of health come before enhancement, and the needs of a people should determine the priorities in their context. "The risks of genomics and genetics could be greater if they are introduced in a precarious public health context, where basic health services are underdeveloped and marked by huge inequalities in access."[38] But maybe here the ethical challenge is not only to moderate the race for enhancement, or to ask questions about allocation of resources, or about distribution of benefits, or even about the risks of hindering the evolution of the genetic chain. The obsession for pragmatic and efficient perfection can threaten the heart of human dignity. In the race for genetic perfection we risk losing our capacity to love, to accept limits, to welcome neighbors in their limitations. It would be a new version of intolerance in times of genomics. It is well to read again John Locke[39] and the classic criticisms of political and religious intolerance in order to learn genetic tolerance for today.

Practices of Ethics in Genomics: Learning a Social Construction

Between reflections on ethics, its applications, and ethical practices, we very often find some tension between idealism and realism.

Such tension takes on a special color in the developing countries, where the social inequalities are bigger. How can ethics proceed in this context? Theology in Latin America has called attention to a close interaction between theory and practice.[40] In that interaction, ethical practices affect the quality of ethical theory. This suggests that it is necessary to take into account the social conditions in order to improve our ethical reflection on human genetics. It will be important to act not only in favor of ethical analysis of certain procedures and therapies but also for a broad education for ethics in genomics, education for foundational values, and for social evaluations as well as practical applications. In fact, ethical reflection and ethical practices are a kind of social construction in which a comprehensive learning process occupies a central place.[41]

Considering the frame of inequalities, needs, and vulnerabilities of the Latin American context, the report of a regional consultation on genomics promoted by WHO in Brasília (July 2001) affirms that

> the meeting put great stress on issues of governance, i.e. the necessity of a framework for application of genomics that will restrain the potential for harm and foster universal access to benefits. This is all the more important since misinformation and greed have already led to inappropriate genetic testing and that this technology presently develops in a context where markets are all-powerful.[42]

This conclusion shows in a direct way the double challenge for a social ethics in this context—to defend ethical practices amid the great vulnerabilities of the population and to empower the people before forces that are not guided by ethical principles.

This same group presents a good synthesis of many questions that are implied in this context:

> for the development of genomics to proceed ethically and for its contribution to health to materialize, several *enabling processes* are needed. These depend on a democratic and informed social dialogue, involving the scientific community, the media, the public, the representatives of civil society and the community at large. For a successful dialogue, a high quality information is essential and this implies responsibilities at all levels, including the medical, scientific, policy and local leaderships. This is also a prerequisite for a sound research policy in the American region. Although their

> research infrastructure varies widely, there is scope for excellence in selected topics in every country. In addition, a close connection between research and the provision of services is the best way to ensure that research investments will yield a return to the public.[43]

In a democratic context, public policies require good participation of civil society for the construction of values and of norms and for the construction of corresponding mechanisms that favor learning and its application. Technical and legal solutions will be insufficient or will have an insufficient quality without the social learning of the values that sustain them. But how can efforts in this direction be strengthened? Comments on three types of concrete initiatives can provide a clearer idea about this.

Social Spaces for Reflection and Academic Debate

Ethics in genomics requires advanced reflection, to be developed with the participation of specialists. In fact, genomics has been challenging the bases of the usual interpretation we have been using to understand ourselves and human life. The conceptual and philosophical revolution it provokes is, in a certain way, prior to and more serious than the specific ethical dilemmas it addresses. It requires, therefore, study and dialogue. In biomedical areas, for a long time legal concerns have prevailed. Over the last ten years in Latin America, concerns about ethical problems have emerged more with the development of bioethics. This is becoming progressively an obligatory discipline in academic programs. The universities are a privileged social space for this reflection because they can provide the necessary variety of specialists and the infrastructure to promote analysis and debate. But the presence of bioethics in the universities is due to the emergence of ethical concerns in the academic community as a whole. Concern with ethics has motivated groups, as well as medical and hospital associations, to promote interesting events. Simultaneously, bioethics societies were created and began to play an important role in the debates. Some universities and other institutions were strongly stimulated to take bioethical questions more seriously. So, these institutions and societies have had in the last decade a role of extreme importance for the Latin American context.[44]

Information and Popular Education in Genomics

Ethics in genomics cannot be reduced to theoretical discussion or to the decisions of specialists. The results of research, many times as mere possibilities or hypotheses, exercise a fascination on public opinion and agitate the social imagination. Basic values in the foundation of the understanding of humanity are touched. This gives rise to a double question about public information and public education. Information of good quality is necessary in order to contemplate the public good and to moderate sensationalism or the "greed of the all-powerful market," as was said above. Information becomes a first step toward education, introducing people to the underlying ethical problems and the values that are at stake. How to do this? Maybe, before speaking of the media, it is important to consider elementary teaching in public and private schools. Schools could provide a network through which to develop an educational process parallel to what is now going on in the universities. Merely technical or biological information is augmented by scholarship and dialogue that promote discovery and awareness of the ethical questions inherent in biotechnological development.

This is a great challenge for the precarious elementary and secondary education system in Latin America in general. "In S.Paulo, the agency that has the main role of producing scientific knowledge in Brazil, the General Office of State for Education, was reducing the workload of the disciplines of sciences. In high school, just one (hour) class of biology is planned in a week; and not a few public schools, for lack of teachers, choose between physics or chemistry, between history or geography."[45] On the other side, initiatives such as the project "Genetics in the Square" try to take the information to the public arena through teachers' groups and popular animators. The question is to follow the information with a good amount of ethical reflection.

The print and television media have an undeniable role in disseminating information, as we all know. But their service in educating the public has been variable and, at times, ambiguous. The forms of provoking the public's participation, exploiting the sensationalism of the news, privilege the emotions of "to be against" or "to be in favor" of a new technology, without fostering discussion of the reasons. This conducts us to another ethical question. Who decides on the information, on the suggestions of values and criteria for the solution of dilemmas, on the cultivation of a global sense in order to humanize biotechnology, and on the whole ideology presented by the media? Those questions are less discussed in the Latin American context. A concrete example could help here. At the end of 2000 in Brazil, the major private TV network of the country was starting a soap opera entitled "The Clone." Specialists in communications estimated this production would command an audience in Brazil of about 40 million spectators during its six months presentation. Like other similar productions, it would be translated into Spanish and transmitted to other Latin American countries, probably reaching another 40 million viewers. Such power to communicate on such important issues for humanity's future is obviously too big to be in the hands of just one private group financed by market interests.

Public Policies, Government Bodies, Legislation

The participation of the developing countries in ethical reflection on genomics depends also on public policy. The problem begins with international relationships in this area. "The rich countries need to act immediately to guarantee that the progress of genomics in the area of health also benefits poor countries. That is a condition necessary to avoid a scenery of exclusion in biotechnology, as it already happens with the technologies of information and agriculture. . . . According to Daar,[46] it is necessary to make the developing countries qualified to create their own bases of research, as already happens in Brazil."[47] This supposes investments driven by public policies, without thereby excluding partnerships with private initiatives that are concerned with the social good. This issue of international social ethics is important for ethics in genomics because it could help to show the great vulnerability of the developing countries in this field. They are exposed to exclusion, as mere consumers of results, or still worse, as mere objects of research.

A second point concerns the public policies of each country. The social ethics of investments, distribution of resources, and protection of vulnerable people and individuals depends a good deal on such policies. "To formulate public policies in this land is an exercise of freedom, when we were already threatened by the difficulty to reflect and to build convictions, tasks that demand time, encounters, dialogue, confrontation of interests and contradictory reasons and an open intellectual debate."[48] In the Latin American context, many official organizations, like national commissions of research ethics, biosafety commissions, sanitary and similar surveillance general offices, are created as a result of a long effort in that social dialogue. Their influence in promoting ethical reflection is mostly indirect, because the whole society is very concerned with legal solutions. But they help at least to keep the ethical problems as social questions. The existence of such examples in developing countries is unfortunately variable and sometimes lacks consistency.

A third point consists of the challenge to have norms and laws that correspond to ethics in human genetics. That is a well-known problem in the world. But in the context of developing countries, the elaboration of laws has a particular challenge in overcoming the pragmatism that bends to the dominant interests of a political and economic type. The vulnerability of the social context as a whole is once more a reality we must deal with. The law schools in the developing countries have been collaborating frequently in ethical discussion about issues regarding which public regulation is needed.[49] Eventually combined with the efforts of bioethics, such discussion results in so-called *biolaw*, which expands juridical foundations through interdisciplinary dialogue. Some specific consequences have already begin to appear.[50]

At the end of these considerations, we can admit that the challenge of the construction of ethics in the development of human genetics is particularly important in the context of the developing countries. But we could observe also that its solution is not restricted to this context. With the development of genomics, we see power growing in our hands but also inequality. The ethical solution demands the humanization of power above all. Humanization means wisdom and balance. In the language of L. Boff:

> Not sufficient is critical reason nor symbolic reason, present in the religions and in the arts, nor emotional reason, underlying the world of the values; nor even the resource of tradition, of common sense and of the wisdom of the people. All those instances are important but none of them is

> enough by itself to guarantee the balance. This demands the articulation of all forces. Balance evokes wisdom, that is precisely the knowledge of fair measure, of the consideration of the "pros" and "cons," a wisdom that has flavor because it picks the best of each thing and of each situation, in a halfway attitude of lack and of abundance.[51]

Power that destroys shows itself fragile and impotent. The vulnerability of the developing countries is a chance for the increasing power in human genetics to become constructive and not threatening. This may bring good news for all humanity.

Notes

1. "Os avanços da ciência no Brasil," *O Estado De Sao Paolo* (editorial) 122, no. 39.177 (February 22, 2001) A3.
2. "Os frutos do Genoma Cana," in Revista Pesquisa FAPESP (Sao Paolo State Research Foundation), available at www.revistapesquisa. fapesp.br. (no. 67 [2001]: 23); "Do Câncer ao Genoma Humano," in Revista Pesquisa FAPESP, available at www.revistapesquisa. fapesp.br. (no. 39 [1999]: 11–15).
3. "Censo científico," in Revista Pesquisa FAPESP, available at www.revistapesquisa. fapesp.br. (no. 62 [2001]: 18–19).
4. Corina Bontempo de Freitas and M. Lobo, "O Sistema CEP/CONEP," in *Cadernos de Ética em Pesquisa* (Brasilia) 4 (2001): 7, 4–13. The projects of human genetics research done in 2000 were 12.8 percent of all projects examined by the Brazilian National Commission for Research Ethics.
5. See Márcio Fabri dos Anjos, "Bioéthique et inégalités socials," in *Théologiques* (Rev. Faculté de Théologie, Univ. Montreal) 7, no. 1 (1999): 19–34; idem, "Power, Ethics and the Poor in Human Genetics Research," in *The Ethics of Genetic Engineering*, ed. Maureen Junker-Kenny and Lisa Sowle Cahill (London/Maryknoll, N.Y.: SCM Press and Orbis Books, 1988), 73–82; idem, "Medical Ethics in the Developing World: A Liberation Theology Perspective," in *Journal of Medicine and Philosophy* 21, no. 6 (December 1996): 629–37.
6. Fernanda Carneiro, "Nosotras: ética e políticas públicas no contexto das culturas da América Latina," in Fernanda Carneiro and Maria Celeste Emerick, *Limite: a ética e o debate jurídico sobre acesso e uso do genoma humano* (Rio de Janeiro: FIOCRUZ, 2000), 113–27.
7. Trajano Sardenberg et al., "Análisis de los aspectos éticos de la investigación en seres humanos contenidos en las intrucciones a los autores

de 139 revistas científicas brasileñas," in *Acta Bioethica* 6, no. 2 (2000): 302–3.

8. Alex Mauron, "Regional Consultation on the ACHR Genomics Report," WHO, Brasilia, July 16–17, 2001, no. 4 (forthcoming by UNESCO).

9. See Luiz Carlos Susin, *Mysterium Creationis* (S. Paulo: Paulinas, 1999); Luiz Carlos Susin, ed., *Sarça Ardente. Teologia na América Latina: Prospectivase* (S. Paulo: Paulinas, 2000).

10. Agostín Lix Costop, *Evangelização e tempos tecnológicos: Análise a partir do universo religioso maia-guatemalteco* (Centro Univ. Assunção, 2001).

11. See Eric Thompson, *Historia y religión de los Mayas* (México: Siglo Veinte, 1980); Sylvanus Morley, *La civilización Maya* (México: Fondo de Cultura Económica, 1975); Miguel Leon Portilla, *Tiempo y realidad en el pensamento Maya* (México: UNAM, 1986).

12. Gustavo Gutierrez, "Situación y tareas de la teología de la Liberación," in *Alternativas* (Nicaragua: Lascasiana) 8, no. 18–19 (2001): 53– 73.

13. Leonardo Boff, "O pobre, a nova cosmologia e a libertação," in *Sarça Ardente: Teologia na América Latina: prospectivas*, ed. Luiz Carlos Susin (S. Paulo: Paulinas, 2000), 195.

14. Leonardo Boff, "Paz como equilíbrio do movimento," *Folha de Sao Paulo* 81, no. 26.474 (September 26, 2001): A3

15. Alwin Toffler, *Powershift: As mudanças do poder* (Rio: Record, 1990): 495–98.

16. There are different meanings to this term. See William F. Arndt and F. Wilbur Gingrich, *A Greek-English Lexicon of the New Testament and Other Early Christian Literature* (Cambridge: University Press, 1957), 277–78.

17. David Noel Freedman, ed., *The Anchor Bible Dictionary* (New York: Doubleday, 1992), 5:444–45.

18. This term is used 108 times in the New Testament. See William F. Moulton and Alfred S. Geden, eds., *A Concordance to the Greek Testament according to the Texts of Wescott and Hort, Tischendorf and the English Revisers*, 5th ed. (Edinburgh: T. & T. Clark, 1978), 347–48.

19. *Dynamis kai exousia* (δύναμιν καὶ ἐξουσίαν).

20. Many times this term appears as an adjective (*dynatos*) and indicates that Jesus was "*powerful* in actions and in words before God and before all the people" (Luke 24:19). For other uses of this term, see Arndt and Gingrich, *Greek-English Lexicon*, 206–7.

21. See Arndt and Gingrich, *A Greek-English Lexicon*, 755–56; and John L. Mackenzie, "Miracle," in *Dictionary of the Bible* (New York: Macmillan, 1978), 612–13.

22. See Calixto Vendrame, *A cura dos doentes na Bíblia* (S. Paulo: Loyola, 2001).

23. See Arndt and Gingrich, *Greek-English Lexicon*, 450; see also the verb κρατέω, in *A Greek-English Lexicon*, 449.

24. See Mark 10:41–45; Matthew 20:24–28; Luke 22:24–27.

25. See James F. Keenan, "Whose Perfection Is It Anyway?: A Virtuous Consideration of Enhancement," *Christian Bioethics* 5 (1999): 104–20.

26. For a general introduction, see A. Fonck, "Perfection chrétienne," in *Dictionnaire de Théologie Catholique* (Paris: Letouzey et Ané, 1933), 12/I:1219–51; John Passmore, *The Perfectibility of Man* (London: G. Dockworth, 1970).

27. Franz Georg Untergassmair, "Vollkommenheit," in *Lexikon für Theologie und Kirche*, ed. Michael Buchberger and Walter Kasper, 3d ed. (Freiburg im Breisgau: Herder, 2001), 10:876. We consider especially the Hebrew adjective *tamin.*

28. Franz Georg Untergassmair, "Vollkommenheit," 876.

29. Arndt and Gingrich, *Greek-English Lexicon*, 816–17 (τέλειος in Greek).

30. See Theodor Seidl, "Reinheit," in *Lexikon für Theologie und Kirche*, ed. Michael Buchberger and Walter Kasper (Freiburg im Breisgau: Herder, 2001), 8:1011–12.

31. John 9: 3 (ἵνα φανερωθῇ τὰ ἔργα τοῦ Θεοῦ ἐν αὐτῷ).

32. Matthew 5:48 (τέλειοι ὡς . . . τέλειός ἐστιν).

33. Matthew 19:21 (εἰ θέλεις τέλειος εἶναι).

34. Luke 6:36.

35. See Jon Sobrino, *El principio misericordia* (San Salvador: UCA, 1993).

36. See Richard Tarnas, *A epopéia do pensamento ocidental, para compreender as idéias que moldaram nossa visão de mundo* (Rio de Janeiro: Bertrand Brasil, 1999).

37. Jean Bartoli, *O ideal de perfeição apresentado aos executivos na revista Exame: um discurso religioso sob a linguagem técnica do management?* (S. Paulo: PUC-SP, 2001).

38. Alex Mauron, "Regional Consultation on the ACHR Genomics Report" (forthcoming by UNESCO).

39. John Locke, *Lettre sur la tolerance* (Paris: Quadrige-PUF, 1995); John Locke, *Lettre sur la Tolérance et autres textes* (Paris: Flammarion, 1992).

40. Clodovis Boff, *Teoria e Prática: Teologia do político e suas mediações* (Petrópolis: Vozes, 1993); idem, *Teoria do Método teológico* (Petrópolis: Vozes, 1998); Enriqe Dussel, *Ética de la liberación en la edad de la globalización y de la exclusión* (Madrid: Trotta, 1998).

41. See International Bioethics Committee of UNESCO (IBC), "Round Table on 'Education on Bioethics,'" *Proceedings*, Seventh Session (November 2000), 2:3–29.

42. Alex Maroun, “Introduction,” *Regional Consultation on the ACHR Genomics Report* (forthcoming by UNESCO).

43. Alex Mauron, “Introduction,” in *Regional Consultation on the ACHR Genomics Report.*

44. See Márcio Fabri dos Anjos, “Notes on Bioethics in Brazil,” *Biomedical Ethics* 5, no. 1 (2000): 42–45.

45. R. H. Bellinghini, “Professores aprendem a ensinar com a genética,” *O Estado De Sao Paolo* 122, no. 39437 (October 8, 2001), A16.

46. Abdallah Daar and Peter Singer, “Harnessing Genomics and Biotechnology to Improve Global Health Equity,” *Science* 294, no. 5540 (October 5, 2001): 87–89. In the abstract, the authors say: “But how can genomics be systematically harnessed to benefit health in developing countries? We propose a five-point strategy, including research, capacity strengthening, consensus building, public engagement, and an investment fund.”

47. Henrique Escobar, “Avanços ainda estão distantes de países pobres,” *O Estado De San Paolo* 122, no. 39437 (October 8, 2001), A15.

48. Fernanda Carneiro and Maria Celeste Emerick, eds., *Limite: a ética e o debate jurídico sobre acesso e uso do genoma humano* (Rio de Janeiro: FIOCRUZ, 2000), 9.

49. See Adriana Diaferia, “Princípios estruturadores do direito à proteção do patrimônio genético humano e as inoformações genéticas contidas no genoma humano como bens de interesses difusos,” in *Limite*, ed. Carneiro Emerick, 167–84.

50. Enrique Varsi, *Derecho genético: principios generales*, 4th ed. (Lima: Grijley, 2001), 548. See also www.viajuridica.com.pe/index .

51. Leonardo Boff, “Paz como equilíbrio do movimento,” A3.

PART II

INTERDISCIPLINARY RESPONSES

8

Genetics, Theology, and Ethics: Toward Complexity?

BARTHA MARIA KNOPPERS

If we were to search for a common thread or characteristic of this book, it would be its inherent complexity. Complexity not in an obtuse or chaotic sense but rather that of modern physics, which understands complex systems to be dynamic, nonhierarchical, and epigenetic. The subject matter lends itself to this of course but in and of itself cannot explain the creative and compelling nature of the arguments and discussion found therein.

Indeed, labels ranging from conservative to liberal, from political to religious, do not really fit this book or any of the chapters. Three intertwining themes with different proposed solutions can serve perhaps to illustrate this extremely rich and valuable attempt to reconcile different disciplines as well as approaches within disciplines. The first is that of the role of science. The second is that of the responsible freedom of citizens. The third is that of the social involvement of citizens in society.

Role of Science

Whereas the Faustian alliance of man with Science-Monster is decried by Mieth, Hansen and Schotsmans see the "playing God" analogy as serving to exacerbate the DNA mystique or DNA as icon. FitzGerald examines the issues of how scientific discoveries affect what is "natural" or "normal." How are human physiology and behavior to be delineated? Two other authors (Anjos and Cahill) move from possible individual or social abuses of scientific prowess to the relationship between our understanding and response to the challenges of the collective genome. For these authors, at the level of humanity, the shared gene pool or common heritage is seen as a reason for focusing on the global health goals of science as a priority for direction. Whatever the point of departure, mistrust or trust, individual or global, none of the authors would leave either the freedom of research or that of citizens totally untrammeled.

Responsible Citizens

While the notion of gift is not commonly used in this book, there is no doubt that scientific advances are seen as the expression of the gift of human intellect and inventiveness. The range of expression is, however, to be limited by the responsibility inherent in co-creation ensuing from the creator (Mieth) or co-creatorship (Hansen and Schotsmans). This freedom in responsibility allows the constant integration of new scientific information in a dynamic way (FitzGerald). It is not only the presence of moral virtues and ideals (Keenan) that guides our decisions as to how we will shape ourselves but also the recognition of our finiteness (Mieth) and personal as well as global imperfections (Anjos). Responsibility and appreciation of the gift of freedom lead to improvement through relationships and social actions.

Relational and Social Involvement

Human self-determination is only possible within the context of interpersonal community (FitzGerald). It is relationships and freedom, not substance and reason (FitzGerald), that allow us to continually shape ourselves (Keenan). It is the potential for and the development of a dynamic relational ethics that makes possible our creative powers. This creativity or "creatorship" requires the integration of all disciplines as equal (FitzGerald).

Called either "praxis" (Mieth), "communicative power in participatory construction" (Anjos), or the fostering of the "common good" (Cahill), the call to power sharing based on solidarity is more than a recurrent leitmotiv. It is the nexus of the vision of humanity shared by all the authors. The social nature of science compels us to serve the promotion of the human person within a just society.

Conclusion

Genetics is a very particular area of science—one that culturally touches citizens and communities "in the shrine of their symbols" (Anjos). Therefore, it calls for a complex systems analysis in the framing of our social and environmental approaches. To do otherwise makes it a threat (Anjos) and not constructive of the citizen within society. It is not a question of how quickly we direct, adapt, adopt, and integrate but how well (Fitzgerald). If technical possibilities precede the legally permissible (Mieth), it is perhaps because we focus on legal tools (e.g., criminal prohibitions, patents, etc.) as simplistic "quick-fixes" or for political assuagement. As a law professor, I am more convinced of the power of the ideas in this book than a preemptive law or an ad hoc legal response. Law is at a level of minimum consensus necessary to maintain the social contract. Its complexity and challenge lie within the continuing debate on the proper role of science, on our freedom and responsibility, and on the possibilities for relational and social empowerment. Classical monogenic, polarized debates couched in polemic ruin true exchange, harm constructive relationships, and halt or kill the potential of the law for responsible social change. Science as freedom, science as responsibility, and science as a tool for global justice are all possible, if we can expose the ideas found in this book.

9

Human Genetic Research Today and Tomorrow: Reflecting on Ethical and Scientific Challenges

ANDREA VICINI, S.J.

In the following pages I propose a few reflections at the intersection between ethics and science inspired by this volume's major contributions. My aim is to be part of the conversation stimulated by the authors' insights and by their ethical approaches as well as by the questioning of researchers working in human genetics.

Four Ethical Remarks

First, in their diversity, the major contributions that we find in this volume highlight a common ethical choice: they propose an *overall and comprehensive ethical approach*. In such a way, they avoid limiting their ethical proposal to identifying the single ethical issues that characterize the current techniques or future possible developments and applications in human genetics, and to suggesting specific ethical tools to address each one of those issues. Hence, our authors invite us to examine the ethical issues raised by progress in genetics as part of concerns related to health care and scientific research today. At the same time, they propose helpful ethical resources to address these major concerns.

Second, together with the authors' choice of proposing an overall ethical approach, they find unsatisfactory an individualistic ethical approach in dealing with the issues raised by today's progress in genetic technology. This critical *common ground* is significant, particularly when we remark that these authors come from different cultural backgrounds and operate in diverse social contexts (Europe, North America, and South America). Such a commonality is neither evident nor largely present within the current ethical debate. At the same time, their diversity of approach is not eliminated. It appears in the ethical resources that they call to our attention and that they use to articulate their overall ethical proposal.[1] This diversity, however, does not indicate incompatibility, but it points toward an approach, both dialogical and conversational, that favors us in joining the debate. The better understanding of the issues at stake, and of the most helpful ethical approach, is not the prerogative of a scholar smarter than the others, but it depends on a clarifying effort that requires open and well-documented debate among ethicists, as well as among scientists, in society.

Third, by focusing on theological contributions, the relevance of religious discourse and rationales within the public arena on issues concerning health, research, and society becomes evident.[2] Ethical insights and suggestions can be theologically grounded, even when they concern political choices in research. They can aim to be part of the current conversation within society on the possibilities and dangers of genetic technology. Our authors show how this debate is promoted by equality, reciprocity, and by being respectful of the often contrasting arguments and proposals in the debate.[3]

Finally, to consider the authors' contributions as a whole, they propose that we address the complexity of the issues concerning today's genetics by using a multiple, synergic, theologically articulated, and interdisciplinary approach. I share this choice made by the authors, and I find insightful their proposal. At the same time, despite the fact that more and more we are living in a global world, how do we account for particularity—culturally, socially, politically, and religiously? How do we settle the tension between, on the one hand, an approach that is universally ethically sound and, on the other hand, issues of particularity, which depend on the cultural diversity and which characterize our social and political contexts (see Fabri)?

Today's genetic technology is developing in rich industrialized countries, both as research and as diagnostic tools in human health (i.e., genetic testing and genetic screening). But, in the future, this trend could change, depending on the availability of cheaper genetic technology, on the access to genetic databases via the Internet, and on support in terms of formation given to scientists in less developed countries. Hence, if my hypothesis is correct, in a few years time it will be interesting to see whether the ethical concerns and approaches proposed by our authors, with their universal scope, will still be relevant, or whether we will need also to emphasize particularity. In such a case, particularity should not diminish the importance of universal ethical approaches but would represent a further step that is necessary to articulate (neither to dismiss nor to challenge) a universal approach in concrete contexts.

Curiously, this could confirm what genetic technology seems to highlight today. On the one hand, the mapping and sequencing of the human genome reveals the high degree of genetic commonality among human beings and, to a certain extent, even with certain animals (i.e., comparative genomics). On the other hand, the increasing genetic knowledge gives us insights into the local, geographical, and familial character of specific genetic mutations, even insofar as the genetic component of the same disease can differ locally.[4] Genetically, the better understanding of what is universal could lead us to discover new types of particularity.[5]

Between Science and Ethics

The authors' suggestions and insights lead me to pay attention to current scientific developments in genetic research as an indication of current and possible future scientific and ethical challenges. I propose two examples that seem to indicate that we should not avoid checking our overall ethical approach with specific technological developments.

First, recent discoveries concerning cell genetic functioning indicate that RNAs[6] intervene in modulating and in regulating the genes' expression.[7] This stresses the need for a *holistic approach*, that is, an understanding of normal and pathological cell functioning, as well as genetic expression, that does not limit itself to knowing the DNA sequence[8] but that aims to promote the cell's normal functioning with the possible related therapeutic applications. In other words, an accurate and detailed, but inclusive, scientific approach appears to be necessary.

This field of research could improve our chances of setting up new protocols in genetic therapy, hopefully with more success than in the case of past and current attempts.[9] At the same time, the research on genetic enhancement and the ethical debate surrounding it could become more focused by promoting what is already present within the cell as a part of its functioning and potential.[10] Finally, the expectations on human reproductive cloning by nuclear transfer could be scientifically challenged because of the complexity of the cell's expression of its genetic heritage and the difficulty of respecting it by simply transferring the cell's nucleus.[11] Hence, scientific developments in genomics could have repercussions for our way of identifying the ethical issues at stake. They could also require that, as ethicists, we support research that aims to offer a better understanding of normal cell functioning. At the same time, the ethical issues that concern justice come to the forefront, particularly when we reflect on the availability to all citizens of the possible therapeutic applications from the expected increase of knowledge in this research field.[12]

Second, by reflecting on global health today, both scientifically and ethically, the many ongoing emergencies (e.g., infectious diseases and food-air-water quality) urge that genetic research focuses on them without depending on immediate economic revenues that could come from research in these fields.[13] A critical approach—as well as the promotion of the common good and the strengthening of virtuous attitudes in scientists, administrators, and politicians—appears to be extremely relevant and urgent to those who care for good living conditions on earth for humankind today as well as for future generations.

Many areas of current and future genetic research (e.g., on stem cells) require from ethicists an accurate scientific understanding, while scientists should be attentive to the possible need for an overall ethical view as well as for specific ethical suggestions and resources.[14] At the same time, I believe that scientific developments could lead us to reaffirm the importance and urgency of addressing wider ethical concerns related to health by understanding it as a common good of humankind and a value that needs to be promoted by virtuous citizens who are attentive to market dynamics. The authors' choice of helping us to reflect on genetics by focusing on overall ethical concerns and by using ethical resources that aim to promote health, human flourishing, the common good, and good living conditions on earth appear to be ethically appropriate.

Third, we could also ask ourselves what researchers want from ethicists and from society at large, and vice versa. Often scientists highlight the need for a sharp understanding of the importance of their research, a fair treatment of their research by the media, sufficient funding, and careful attention given to the possible benefits that could depend on the results of their research. Ethicists would respond that it is also necessary to have a refined perception of the ethical issues at stake. Hence, dialogue, interaction, and collaboration appear to be values—as well as virtues, when we focus on the persons who live them—ethically important for both scientists and ethicists.

My hope is that, in our world, today and in the future, we (without excluding anyone) can experience and enjoy more and more the benefits that can come from these collaborations, both scientific and ethical, in making available some of the expected achievements made possible by progress in genetic technology—for example, genetic tests, genetic screenings, and genetic therapies.

Notes

1. I refer to a critical attitude that unmasks manipulative rhetorical statements and that articulates our anthropological finitude (see Mieth), to a theological understanding of human nature (see FitzGerald) and human agency (see Hansen and Schotsmans), to the role of virtues in shaping moral life (see Keenan), to the relevance of aiming at promoting the common good (see Cahill), to stressing the preferential option for the poor and to strengthening social justice (see Fabri).

2. For a similar attempt, see D. Hollenbach, *The Common Good and Christian Ethics: New Studies in Christian Ethics,* ed. R. Gill (Cambridge, U.K.: Cambridge University Press, 2002).

3. For other recent examples of a theological approach on issues concerning genetics, see R. Song, *Human Genetics: Fabricating the Future, Ethics and Theology* (Cleveland: Pilgrim Press, 2002); T. A. Shannon and J. J. Walter, *The New Genetic Medicine: Theological and Ethical Reflections* (Lanham, Md.: Rowman & Littlefield, 2003); D. H. Smith and C. B. Cohen, eds., *A Christian Response to the New Genetics: Religious, Ethical, and Social Issues* (Lanham, Md.: Rowman & Littlefield, 2003); S. Brooks Thistlethwaite, ed., *Adam, Eve, and the Genome: The Human Genome Project and Theology: Theology and the Sciences* (Minneapolis: Fortress Press, 2003).

4. In other words, the same disease does not always depend on the

same type of mutation (here I am not discussing genetic polymorphisms) but on different genetic mutations in related groups of affected subjects and families.

5. I am grateful to Amalia C. Bruni, M.D., director of the Regional Center for Neurogenetics (Lamezia Terme, Cosenza, Italy) for this insight taken from her current research on neurological diseases with a genetic component and their familial and geographical distribution.

6. At the end of 2002 and 2003, the prestigious scientific journal *Science* affirmed that the recent discoveries concerning the various types of RNAs and their functioning are among the most relevant current scientific breakthrough. See J. Couzin, "Small RNAs Make Big Splash," *Science* 298, no. 5602 (2002): 2296–97; D. Kennedy, "Breakthrough of the Year," *Science* 298, no. 5602 (2002): 2283; News and Editorial Staffs, "The Runners-Up," *Science* 302, no. 5653 (2003): 2239–45, here 2041.

7. These recent discoveries concern (1) small temporal RNAs, called *micro RNA* (miRNA). They control gene expression by suppressing translation or by degrading the targeted messenger RNAs (mRNAs); (2) *interfering RNAs* (RNAi's). They intervene in gene silencing and are produced from aberrant RNA (probably double-stranded) by the enzyme Dicer. RNAi's can also modify the structure of chromatin and change gene expression as well as the whole genome expression. See J. Couzin, "Small RNAs Make Big Splash"; D. Kennedy, "Breakthrough of the Year."

8. On the complexity of cell functioning, see current studies concerning epigenetics. The term indicates "changes in gene expression that persist across at least one generation but are not caused by changes in DNA code"; Couzin, "Small RNAs Make Big Splash," 2296. On the possible role of miRNAs in epigenetic phenomena, see C. Dennis, "Altered States," *Nature* 421, no. 6924 (2003): 686–88; idem, "The Genome's Guiding Hand?" *Nature* 420, no. 6917 (2002): 732. See also S. I. S. Grewal and D. Moazed, "Heterochromatin and Epigenetic Control of Gene Expression," *Science* 301, no. 5634 (2003): 798–802.

9. On genetic therapy and its complications, see S. Hacein-Bey-Alina et al., "A Serious Adverse Event after Successful Gene Therapy for X-Linked Severe Combined Immunodeficiency," *New England Journal of Medicine* 348 (2003): 255–56; P. Noguchi, "Risks and Benefits of Gene Therapy," *New England Journal of Medicine* 348 (2003): 193–94.

10. On genetic therapy and enhancement, see L. Walters and J. Gage Palmer, *The Ethics of Human Gene Therapy* (New York/Oxford: Oxford University Press, 1997); A. R. Chapman, "Unprecedented Choices: Religious Ethics at the Frontiers of Genetic Science," in *Genetics and the Sciences,* ed. K. J. Sharpe (Minneapolis: Fortress Press, 1999).

11. Various authors have highlighted problems in animal reproductive cloning concerning the animal development and metabolic state. See, as an example, G. Kolata, "Sheep Clone's Cells Aging Faster Than She Is," *New*

York Times, May 27, 1999, A24; "Multiple Gene Defects Found in Clones," *Science* 297, no. 5589 (2002): 1991.

12. Pharmacogenomics is another example of a rapidly developing research field in which potential benefits in terms of treating and healing diseases in subjects responding, or not, to drugs cannot be separated from the expected major economic profits of labs and companies involved in this research field.

13. Research on malaria is a well-known example. Despite the extremely relevant benefits in terms of human health and economic repercussions globally that could be achieved, the expected limited economic advantages from patents for labs and companies funding research in this field do not motivate major funding of this research. The revenues appear to be more limited than in studying diseases that mostly affect the rich industrialized countries.

14. As an example, in the case of studies in population genomics, whenever the research focuses on populations affected by diseases that are associated with psychiatric stigma, as in the case of dementia, the citizens' involvement and the publicity given to the research project cannot be the same as in the case of other research studies on common diseases (e.g., cardiovascular diseases), which normally are not associated with the risk of discrimination and stigmatization. Hence, population genomic studies that aim to highlight the genetic component in psychiatric diseases need a specific ethical approach that preserves the confidentiality of the involved targeted populations.

10

Comments from the Praxis of Predictive Testing, Prenatal Testing, and Preimplantation Genetic Diagnosis for Late-Onset Neurogenetic Disease: The Example of Huntington's Disease

GERRY EVERS-KIEBOOMS

Introduction

This response to the project contributions in the first part of the present volume is written from a very specific perspective, namely, from the praxis of predictive and prenatal testing for late-onset neurodegenerative diseases within a multidisciplinary counseling context. Hereby special attention is paid to the fear of at-risk persons of transmitting late-onset disease to the next generation(s). Huntington's disease (HD) is used as an example. It is, indeed, the first serious late-onset disease for which predictive DNA-testing became possible about twenty years ago, a few years after discovering that the Huntington gene was localized on the short arm of chromosome 4.

Huntington's disease is a currently untreatable progressive neuropsychiatric disorder, characterized by involuntary movements, neuropsychological defects, and personality changes. The mean age at onset is about forty years. Symptoms progress slowly, with death occurring an average of fifteen years after the onset of the disease. Huntington's disease is inherited as an autosomal dominant trait. In 1993, the Huntington gene, containing an expanded and unstable trinucleotide repeat (CAG) in HD patients, was isolated.[1] The availability of predictive tests and prenatal tests, initially by DNA-linkage and since 1993 by direct mutation analysis, was a new challenge for families and professionals. In the meantime, thousands of predictive tests and hundreds of prenatal tests have been performed all over the world. Preimplantation genetic diagnosis (PGD) for HD has become available more recently and is only available in a limited number of centers in the world. We refer to Geraedts and Liebaers for more information on the technical procedure and on the experience gained so far.[2]

Predictive tests for late-onset neurogenetic diseases provide information about an asymptomatic person's *future* health status regarding a specific disease. The test results lead to a new status or psychological identity. An individual prediction about the exact age at onset, the specific symptoms, or the course of the disease is impossible. In other words, a degree of uncertainty persists. Predictive tests for late-onset neurogenetic diseases have far reaching implications for the test applicants, their families, and for society.[3] It is of the utmost importance that a real choice about having or not having a predictive test is safeguarded—a well-informed and free decision of the test applicant without pressure from third parties.

Prenatal testing (full prenatal diagnosis, prenatal exclusion testing, and preimplantation genetic diagnosis) should always be considered within the context of the complex process of reproductive decision making that people at increased genetic risk for a late-onset neurogenetic disease are facing. Moreover, it should be carried out with adequate counseling, preferably in centers with vast experience in predictive testing. A very difficult situation may arise when parents decide to continue the pregnancy wherein prenatal diagnosis reveals a Huntington mutation in the fetus. Indeed, the child to be born may later in his life be harmed by this information, for example, by spoiling, overprotecting, or neglect by the parents. Moreover in this situation the child's own right not to know is already violated before his birth. All these issues should receive attention during counseling before the parents engage in a pregnancy with prenatal diagnosis.

The need for a multidisciplinary approach to predictive testing for HD is the topic of the second section. Third, the uptake for predictive and prenatal testing as well as the factors involved in decision making to have or not have a test for HD are briefly reviewed.

Some major findings of a recent European collaborative study are also presented. Finally, some of the concepts discussed in the first part of this volume will be considered from the perspective of the praxis of predictive testing, prenatal testing, and preimplantation genetic diagnosis for late-onset neurogenetic diseases.

Predictive Testing for Huntington's Disease: The Elaboration of Careful Protocols and the Need for a Multidisciplinary Approach

A large amount of international debate and consultation preceded the implementation of the first predictive and prenatal tests for HD in clinical practice. The guidelines elaborated by the International Huntington Association and the World Federation of Neurology have in the meantime been used as a model for predictive testing for other neurogenetic diseases.[4] They are aimed at protecting test applicants and at assisting clinicians, geneticists, and ethical committees as well as lay organizations to resolve difficulties arising from the application of the test.

Predictive test requests are often approached by a multidisciplinary team consisting of a clinical geneticist, a psychologist, a neurologist and/or a social worker or genetic nurse. The number of professionals involved may differ from one center to another. During the pretest counseling sessions, full information is provided on HD and on the predictive test. The role and psychological meaning of the disease and the test in the course of the life of the testee are explored. The main aim of the pretest counseling sessions is to help people to use sufficient time for reflection, to develop a scenario of their life after a favorable test result or after an unfavorable test result or without having a predictive test, and to make a free informed decision about having or not having a predictive test. After the disclosure of the predictive test result, short- and long-term emotional and social support is systematically provided during follow-up counseling sessions. The partner of the test applicant is encouraged to participate in all pretest and posttest counseling sessions.

Uptake for Predictive Testing and Prenatal Testing for Huntington's Disease and Impact of the Predictive Test Result on Subsequent Reproduction

All over the world the proportion of persons who applied for predictive testing is smaller than could be expected based on intentions and attitudes before the availability of the test: the uptake rate varies between 5 percent and 20 percent. It is obvious that sociodemographic variables hardly play any part in the decision to be tested. Personality profile and individual coping style seem to be the key factors. The motivation to have or not have a predictive test is very complex: there often are major and minor reasons and moreover there is a mixture of conscious and unconscious motives. It is important to keep in mind that in most surveys or other studies only the conscious motives expressed or indicated by the persons at risk for HD are taken into account. In most centers over the world that reported their data, the two major motives for requesting predictive testing are "certainty for the own future" and "family planning."[5] Two articles by Decruyenaere et al. provide a recent evaluation of the psychological distress in the five-year period after predictive testing for HD and a description of the impact of the test result on the posttest partner relationship.[6]

With respect to family planning, the findings in many centers in the world illustrate that carriers of the HD mutation with a desire for children are confronted with difficult new decisions and an additional emotional burden.[7] They have the following options: refraining from children, taking the risk of having a child with the Huntington mutation, using prenatal diagnosis, artificial insemination with donor sperm, in vitro fertilization (IVF) with donor eggs, and more recently, in a small number of centers in the world, preimplantation genetic diagnosis.

When considering prenatal testing for late-onset disease, it is important to make a distinction between "full prenatal testing" aimed at obtaining all available information concerning the specific late-onset disease (before 1993 by DNA-linkage, thereafter by direct mutation analysis) and "prenatal exclusion testing" aimed at determining whether the fetus has the same genetic risk as the asymptomatic at-risk parent belonging to the Huntington family or no risk for HD. In the latter case, one determines whether the fetus inherited a chromosome 4 from the affected grandparent or from the partner of the affected grandparent (no information is given about the genetic status of the at-risk parent), and in principle parents opt for a termination of pregnancy when the fetus has the same risk as the prospective parent. Initially, prenatal exclusion testing was mainly introduced as an option for couples with an uninformative family structure, for whom predictive testing by DNA-linkage was impossible because of technical limitations. After the identification of the gene in 1993, the option continued to be offered to at-risk persons who did not want a predictive test to know their own status but who wanted to prevent the transmission of the disease to their children.

Results of a survey assessing the uptake of prenatal testing for HD worldwide, carried out in close collaboration with the World Federation of the Neurology Research Group on HD (WFNRHD), revealed that the uptake of prenatal testing is low but that the results vary greatly between different countries and between different centers within a country.[8] At country level, detailed data have been published for the Netherlands and the United Kingdom, respectively, by Maat-Kievit et al. and by Simpson and Harper.[9] The prediction of Hayden et al. that the demand for prenatal testing would further decrease in the future because of the optimism about a potential treatment has become true in some countries.[10] In 2001, Hayden reported that not a single prenatal test for HD was performed in British Columbia during the past five years.

To get more insight in the psychological, social, ethical and legal complexity of prenatal testing for HD, a European collaborative multidisciplinary study was carried out in the context of a European Commission funded project (BIOMED-project N° ERB BMH4—CT98-3926). The major objectives of the European project were (1) creating a detailed picture of prenatal testing for HD in six countries in different parts of the European community; (2) getting insight in reproductive decision making of carriers of the Huntington mutation as compared to noncarriers; (3) analyzing the ethical and legal implications of prenatal testing for HD; and (4) stimulating a discussion about the findings between professionals from different disciplines (medical, ethical, psychological, and legal) and HD associations. In November 2000, the results of the European project were presented during a meeting in Leuven, Belgium, and this meeting was the starting point for producing a book with a much broader scope—an extension to other late-onset neurogenetic diseases and to North America. Multidisciplinarity is a major characteristic of this book, edited by G. Evers-Kiebooms, M. Zoeteweij, and P. Harper, and special attention has been paid to the family's perspective, not only in a contribution by the president of the International Huntington Association but also by including several vivid case presentations about prenatal testing and preimplantation genetic diagnosis.[11] A few major findings of the European collaborative study related to the first two aims of the project will be given.

The six European countries (with a total population of about 200 million people) involved in the collaborative study reported 305 prenatal tests for HD in the period from 1993 to 1998.[12] The uptake for prenatal testing was low, almost ten times smaller than the uptake for predictive testing in the same countries. Nevertheless, there were large differences. The countries fall into two groups with the uptake in the United Kingdom, the Netherlands, and Belgium an order of magnitude higher than in France, Greece, and Italy. Almost two-thirds of the total series of prenatal tests were exclusion tests. Despite the availability of full prenatal testing, exclusion testing is still requested by a considerable group of individuals at risk who do not want to know their own status; this is particularly true for the United Kingdom and for Italy. It was rather exceptional that pregnancies wherein the fetus carried the Huntington mutation or wherein exclusion testing revealed a 50 percent risk in the fetus were not terminated. The study also illustrated that about one-third of the couples had repeated prenatal testing.

The European collaborative study clearly revealed a measurable impact of the predictive test result on subsequent reproduction in a large group of test applicants who were not older than forty-five years when they applied for predictive testing: 14 percent of the total group of carriers had one or more subsequent pregnancies versus 28 percent of the total group of noncarriers.[13] A prenatal test was carried out in about two-thirds of the pregnancies in the carrier group. A more refined analysis in the subgroup that is most adequate for this type of evaluation, namely, the group who reported "family planning" as a motive to apply for predictive testing in the pretest period and with a follow-up interval of at least three years, revealed a more pronounced effect: 39 percent of the carriers of the

Huntington mutation had subsequent pregnancies versus 69 percent of the noncarriers. Notwithstanding a desire for children and notwithstanding the presence of "family planning" as one of the major reasons to apply for predictive testing, the majority of carriers had no subsequent pregnancy. It is clear from the above findings that predictive DNA-testing helps a group of couples to fulfill their desire to have children without any risk of transmitting the mutation. In considering all these findings it is important to keep in mind that only a minority of persons at risk for HD make use of predictive testing and prenatal testing. Nevertheless, although systematic research on the nontested group is lacking, all professionals confronted with families with Huntington's disease have experienced in practice that reproductive decision making in this situation of increased genetic risk usually is and remains a difficult dilemma. Whatever option is eventually chosen by a couple at increased risk of transmitting the Huntington mutation, it is of the utmost importance that professionals fully respect this decision and support the couple.

Comments on the Contributions of the First Part of the Volume from the Perspective of the Praxis of Predictive Testing, Prenatal Testing, and Preimplantation Genetic Diagnosis for Huntington's Disease

The following statements in Cahill's introduction and their elaboration by several contributors are the starting point for my comments:

- Catholic tradition provides resources for looking at genetics in terms of the value of scientific progress and healing, of individual decisions focused on reproduction, and on distributive justice and the common good.
- It is less well known that the church and John Paul II specifically have taken a positive attitude toward genetic research in general, as long as it aims to promote human welfare within the limits required by "full respect for man's dignity and freedom."

Genetic research has resulted in genetic diagnostic possibilities in asymptomatic adults (predictive testing), in fetuses (prenatal testing), and in embryos (preimplantation genetic diagnosis), the latter, of course, in close relationship with the achievements for in vitro fertilization. In particular, the last possibility is only available nowadays because some centers in the world were involved in embryo research in the hope that this would lead to help infertile couples to fulfill their desire for children. Because of moral and religious considerations, other researchers, with a comparable hope to help couples in this situation, were more reluctant to engage in this type of research. Nowadays many professionals as well as families with late-onset disease view the availability of preimplantation genetic diagnosis as a valuable option for persons who know that they carry the Huntington mutation and who want to have children without a risk of transmitting the disease to their offspring. As opposed to prenatal diagnosis, it helps them to avoid the confrontation with the emotionally and morally laden decision to terminate a pregnancy, in particular, in a situation of a strong desire to have a child. The psychological and financial cost of the in vitro fertilization procedure, partly due to the low "success rate," is still a problem that can only be reduced at the technical level by improving the outcome of IVF . . . implying additional research on embryos be it in circumstances safeguarding respect for human life. Both prenatal testing and PGD are important options in the context of free reproductive decision making of persons who have to live with a very high genetic risk as well as with the burden of a serious disease in one of their parents and other relatives (the prefiguration of a possible own future) and who want to fulfill their desire for "HD-free" children. They need all our respect, support, and compassion.

Genetic research can promote human welfare within the limits of respect for human dignity and freedom. At present, prenatal testing and PGD are important options for prospective parents, but it is obvious that they can never be imposed by third parties and they should always be preceded by adequate counseling hereby respecting the counselees' own values. Complementary to these achievements, stem cell research and therapeutic cloning—within carefully defined limits—seem important because they may lead to better treatment of patients with Huntington's disease. As such they give hope for the future and this is extremely important for families confronted with Huntington's disease and other currently untreatable late-onset diseases.

The concepts of "common good," "solidarity," "distributive justice" are very relevant when we consider the available genetic services and genetic tests for families with Huntington's disease and other late-onset genetic diseases. At present there certainly is no equal access to these services. In developing countries they usually are not available at all, and, in particular, PGD is only available for a very small minority in some developed countries. In this context, Harper reminds us:

> At present the reproductive issues raised by serious late onset genetic disorders remain a severe burden to those who are affected or at risk, and to their partners and family members. . . . Through prenatal and presymptomatic testing, carried out in the context of skilled and sensitive genetic counseling, we have been able to be of some help to such families, even when they may opt not to pursue the actual testing options. There remain many for whom existing options are unsatisfactory, and probably many more worldwide who never have the opportunity to even consider them. We all need to work towards the situation where the possibilities are not only improved for some families, but also where there is availability for all who need them. Whether families will in the future be better able to be helped towards their goal of achieving their own family free from risk of serious disorder through prenatal or presymptomatic testing, through advances in therapy or through yet other approaches, remains to be seen.[14]

My final comments relate to the collaborative study organized by the Hereditary Disease Foundation resulting in the localization and later isolation of the Huntington gene. The exceptionally large Venezuelan kindred, with more than one hundred patients with Huntington's disease living in the villages around Lake Maracaibo, was essential in this important genetic research and has also led to a major longitudinal study of the disease.[15] In the context of distributive justice and solidarity, a fair return of benefit to this society should continue to receive major attention.

Notes

1. Huntington's Disease Collaborative Research Group, "A Novel Gene Containing a Trinucleotide Repeat That Is Expanded and Unstable on Huntington's Disease Chromosome," *Cell* 72 (1993): 971–83.

2. J. P. M. Geraedts and I. Liebaers, "Preimplantation Genetic Diagnosis for Huntington's Disease," in *Prenatal Testing for Late Onset Neurogenetic Diseases*, ed. G. Evers-Kiebooms, M. Zoeteweij, and P. Harper (Oxford: BIOS, 2002), 107–18.

3. P. Harper and A. Clarke, *Genetics, Society and Clinical Practice* (Oxford: BIOS, 1997).

4. International Huntington Association and World Federation of Neurology, "Guidelines for the Molecular Genetics Predictive Test in Huntington's Disease," *Neurology* 44 (1994): 1533–36.

5. For a review on the impact of predictive testing on the psychological well-being of testees and their family, we refer to G. Evers-Kiebooms et al., "The Psychological Complexity of Predictive Testing for Late Onset Neurogenetic Diseases and Hereditary Cancers," *Social Science and Medicine* 51 (2000): 831–41.

6. M. Decruyenaere et al., "Psychological Distress in the 5-year Period after Predictive Testing for Huntington's Disease," *European Journal of Human Genetics* 11 (2003): 30–38; M. Decruyenaere et al., "Predictive Testing for Huntington's Disease and the Post-test Partner Relationship," *Clinical Genetics* 65 (2004): 24–31.

7. M. Decruyenaere et al., "Non Participation in Predictive Testing for Huntington's Disease: Individual Decision Making, Personality and Avoidment Behavior in the Family," *European Journal of Human Genetics* 5 (1997): 351–63; A. Tyler et al., "Exclusion Testing in Pregnancy for Huntington's Disease," *Journal of Medical Genetics* 27 (1990): 488–95; S. Adam et al., "Five Year Study of Prenatal Testing for Huntington's Disease: Demand, Attitudes and Psychological Assessment," *Journal of Medical Genetics* 30 (1993): 549–56; J. L. Tolmie et al., "The Prenatal Exclusion Test for Huntington's Disease: Experience in the West of Scotland, 1986–1993," *Journal of Medical Genetics* 32 (1995): 97–101; J. D. Schulman et al., "Preimplantation Genetic Testing for Huntington's Disease and Certain Other Dominantly Inherited Disorders," *Clinical Genetics* 49 (1996): 57–58; G. Evers-Kiebooms et al., "Predictive and Preimplantation Testing for Huntington's Disease and Other Late-onset Dominant Disorders: Not in Conflict but Complementary," *Clinical Genetics* 50 (1996): 275–76.

8. G. Evers-Kiebooms, "Recent Evolutions in Prenatal Testing for Huntington's Disease," Presentation during the 17th International Meeting of the World Federation of Neurology Research Group on Huntington's Disease, Sydney (unpublished, August 1997).

9. A. Maat-Kievit et al., "Experience in Prenatal Testing for Hunting-

ton's Disease in the Netherlands: Procedures, Results and Guidelines (1987–1997)," *Prenatal Diagnosis* 19 (1999): 450–57; S. A. Simpson and P. S. Harper, "Prenatal Testing for Huntington's Disease: Experience within the UK 1994–1998," *Journal of Medical Genetics* 38 (2001): 333–35.

10. M. R. Hayden, M. Bloch, and S. Wiggins, "Psychological Effects of Predictive Testing for Huntington's Disease," *Advances in Neurology* 65 (1995): 201–10.

11. G. Evers-Kiebooms, M. Zoeteweij and P. Harper, eds., *Prenatal Testing for Late-onset Neurogenetic Diseases.*

12. M. Zoeteweij et al., "An Overview of Prenatal Testing for Huntington's Disease in Six European Countries," in *Prenatal Testing for Late-onset Neurogenetic Diseases,* 25–43.

13. G. Evers-Kiebooms et al., "Predictive DNA-testing for Huntington's Disease and Reproductive Decision Making: A European Collaborative Study," *European Journal of Human Genetics* 10 (2002): 167–76.

14. P. Harper, "Prenatal Testing for Late Onset Genetic Disorders: Evidence and Insights from Huntington's Disease," in *Prenatal Testing for Late-onset Neurogenetic Diseases*, 203-14.

15. Idem, *Huntington's Disease* (London: Saunders, 1996).

11

Genetics, Ethics, Theology: A Response from the Developing World

Hasna Begum

Because I am trained in philosophy, I find many of the ideas in this volume stimulating and thought provoking, and, as an advocate for feminism, human rights, and a female citizen of one of the poorest countries, I concur in many of its social objectives as well. I shall try to respond to each of the essays and to offer some further perspectives, based on my own varied experiences gathered from being a mere housewife to a professor of philosophy, from being a teenage mother to a gray-haired woman. I had the opportunity to become a member of a few international associations, which provided me with exposure to the present-day world order and to what is going on in the field of ethics, especially bioethics. But unfortunately, the viewpoints expressed in the interest of the developing world seldom could cross the table to be included in the resolutions. Mere rational thinking, in the absence of direct experience of situations unique to poor countries, cannot possibly comprehend them in their uniqueness. This is possibly the reason for such failures.

The first chapter of part 1, "Stem Cell Research: A Theological Interpretation," by Hansen and Schotsmans, effectively enumerates the problem of the relations between genetics, theology, and ethics. This essay also explains the key aspects of the problem in elaborating the process of both reproductive cloning and therapeutic cloning. It also clarifies the key terminologies used in the essay and in any discussion of the same topic. The main objective of the essay is to make the problem under discussion easy to understand for people who are interested in the problem without having specialized knowledge of either genetics or theology. The tables explain cloning techniques and the crucial difference between the method of nuclear transplantation as involved in therapeutic cloning and the method involved in human reproductive cloning. Hence, as the opening chapter of the book, this essay is most suitably placed.

The standpoint of the authors, in their own words, "is essentially close to this more open and optimistic perspective: the human being as a *created co-creator* is in our opinion not standing up against God, but he is able to realize God's intentions in full responsibility." That is, human beings are able to complete the intention of God and can become complementary to his intended act of creating humans free from sickness, suffering, and also from premature death.

The authors do not encourage reproductive cloning, as Christian theology does not allow humans to create other human life artificially. But they do show justification in favor of therapeutic cloning, which involves human embryonic stem cell research or stem cell research. As the basis of their justification, they cite their interpretation of Christian theology as admitting that life begins in the womb after two weeks of conception. Their argument favoring therapeutic cloning is simple but convincing. Human beings are created by God, who makes them free to choose and capable of choosing between good and evil. But at the same time, they fall victim to sickness, suffering, and premature death. The promise of therapeutic cloning adopted by human beings may save humanity from the above-mentioned unwanted evils. These evils are not included in God's *original intention,* though his created humanity is prone to these undesirable consequences because of God's incomplete creation. Thus, being created themselves by God, human beings could be *co-creator* in completing God's intention regarding human creation. This argument seems plausible and quite convincing to people of Christian faith. To me, however, it seems there is a legitimate question here: How could God, a perfect being, create such an incomplete creation? This is a philosophical question and the argument offered seems to be logically fallible.

It may also be mentioned here that the same line of thought regarding reproductive cloning may be adopted by the Muslims, believers in Islam. According to Islam, life begins in the womb of the mother after six weeks of conception. So, therapeutic cloning may be allowed in Islam as the embryonic stem cell could be collected and utilized till the embryo is six weeks old. And minimizing the sufferings of humankind is given much importance in Islam as a religion. It is also mentioned in the Qur'an and in the Hadith that through serving humanity one can worship Allah the merciful. In case of in vitro fertilization (IVF), it may be safely thought that the same justification would be applicable for using the stem cells for therapeutic cloning.[1]

D. Mieth's essay "The Human Being and the Myth of Progress; or, The Possibilities and Limitations of Finite Freedom" follows the essay by Hansen and Schotsmans. Mieth is very cautious to accept any kind of artificial interventions in reproductive processes in general. He does not see humans as *created co-creator* as Schotsmans and Hansen do. Mieth fears that given the sanctions for such interventions human beings may attempt to control human reproduction by excessive and forceful interventions. There is a sound basis to this fear because of the intrinsic finitude and imperfection of humans. Before giving any sanction to even therapeutic cloning, there is need to give the matter serious thought. With regard to reproductive cloning, this fear is even more pronounced. Mieth's conception of the "human meaning of sexual relations and procreation" is not to be taken seriously as sexual relations among humans secularly may have intrinsic worth and meaning without any reference to procreation. It seems that Mieth is quite right, however, in his observation that reproductive interventions are usually carried out on the bodies of women, and these bodies are not selected randomly but the selection of bodies of women are determined by cultural and economic factors. This is cultural and economic oppression, and the poorer these women are, the more they are oppressed. Thus, the women of the developing and least-developed world are likely to be most used as means to the end of gaining from genetic experiments. Here it is also relevant to mention that the whole process of artificial reproduction technology, both negative (controlling population) and positive (diminishing infertility and bringing about enhancement), involves pressure on women. Hence, if we think objectively, women are the worst sufferers. In Bangladesh, for example, of those who use methods for controlling reproduction, only about 4.6 percent are male. The burden of birth control is almost entirely borne by women.[2]

The essay by Kevin FitzGerald, "The Need for Dynamic and Integrative Vision of the Human for the Ethics of Genetics," is a very penetrating discussion of the implementation of genetic technology to modify human genes in order to combat genetic diseases (and deficiencies). FitzGerald takes recourse to philosophical anthropologies in explaining the ethics of genetics. He fears that genetic technologies may possibly be applied to alter the human genome "in an attempt to effect a fundamental change in human nature." Such an attempt involves the need to perfectly understand human nature by the scientists who have to have a clear knowledge into the further future as to the consequences of such applications. FitzGerald is very much concerned about such applications. He also suggests to resolve the ethical questions regarding "which genetic interventions should be pursued and which should be left unattempted" before giving any sanctions to such interventions.

FitzGerald explains the distinctions between somatic and germ-line interventions and between the purposes of such interventions for therapy and enhancement. He reminds us of the fact that a broad consensus of national and international panels have supported the careful pursuit of somatic cell gene therapy (type 1) and rejected the "pursuit of the other three types of genetic interventions—involving either nontherapeutic genetic changes or changes that can be passed on to future generations," because it "is not yet advisable." He also at the same time fears that this consensus in support of only type 1 interventions may not last much longer. The invention of newer technologies seems to have an inner force toward implementation. FitzGerald thinks it is also not very easy to distinguish between therapy and enhancement and cites an example of such blurring in the variation in susceptibility to HIV infection found in different people. And arguments can be put forward in favor of both views. It is alarmingly true that this blurring may make therapeutic interventions in many cases controversial. FitzGerald also warned against germ-line intervention which could potentially be used for horrible abuses and possible tragic disasters affecting many people in the future. The author does not seem to have any objection against carefully thought-out therapeutic somatic cell intervention as it directly affects only the treated individuals. It should be mentioned here, however, that those who cannot afford the interventions would still have to go on suffering. This certainly is a kind of injustice and implies discrimination on the basis of economic status. Economic status, thus, has a bearing on health care. Therefore, there is enough justification for finding a strong relation between economic conditions, health care, and health care ethics.[3]

James Keenan, in his essay "What Does Virtue Ethics Bring to Genetics?," relates virtue ethics to our response to human research in genetics regarding enhancement. He is typically critical of contemporary *secular* attitudes to bioethics that ignore or disregard their "religious companions" and exclude "prophetic and the narrative" discourse. Keenan rightly observes that human identity is historically dynamic, and the virtues that are possessed by human beings are also dynamic. He says, "We engage our understandings of both who we are as humans and who we can virtuously become through the development of history." Through human genetic research for enhancement we may in the future become "virtuously" what we are not in the present. This fact itself involves us (human beings) seriously with a critical understanding of what is going on in the field of human research in genetics, especially with the goal of enhancement. For Keenan, the specific cardinal virtues for contemporary anthropology are basically three in number, of which each is absolutely independent of the other. These three are justice, fidelity, and self-care. The first is generally related to all; the second is specifically related to a few or some; and the third is related to only one, that is, to the self. There is yet another virtue that integrates these three virtues into our relationships not only in the present but also further into the future. It seems these virtues are dynamic for Keenan in the sense that they can vary in degree in a particular individual in the course of time.

All the above contentions of Keenan seem to refer to his view that genetics must be ethically judged from the criterion of virtue ethics with the cardinal virtues he mentions. He says, referring to Paul Lauritzen, "A relational anthropology required us . . . to rethink our understanding not only of the self but of morality in general and of the cardinal virtues in particular."

Márcio Fabri dos Anjos, in his essay "Power and Vulnerability," writes from the viewpoint of a resident in a developing country, Brazil. He puts forward a very relevant point regarding human research in genetics: it is not clear who will control the power and also benefit from it. This is a very legitimate question, indeed, and I do agree with him. I am much concerned about the future of poor countries, which, because of economic oppression, are globally the chess pieces in the hands of rich countries. What effect will genetic research have on them? Bioethicists in general, however, fear that the poor and the poor countries will remain vulnerable in the hands of the rich and rich countries in the age of genomic development. Fabri opines that in Latin America (and in many other developing countries) many social groups will become vulnerable and lack the frame of mind to reap benefit from genetic research. They are also not capable of negotiating with those who have power and thus will fall prey to the rich, who will control genomics. He takes recourse in liberation theology and the Bible for salvation. His view is that power will act as a dynamic force for the common good, which includes the good of the poor, if the teachings in the Bible are followed by people in general. With all his best intentions, Fabri sounds highly idealistic and also, to say the least, too dogmatically sure of the influence of Jesus' concern for curing humans from imperfection, sin, and suffering. He cites "human dignity," "civil society," and other such high-sounding phrases to establish his idealistic and unrealistic opinion. Unfortunately, in reality, the poor are lacking in education and wealth and consequently are bound to become vulnerable in the age of genomics and "globalization."

Lisa Sowle Cahill, in her essay "Genetics, Theology, Common Good," places "the ethics of genetics in the context of the common good." She sounds cautious of "any undue influence of market forces in determining social relations and access to knowledge and goods." Cahill opines that some international policy statements place genetic knowledge in the context of social responsibility and thus play a positive role toward common good. But she also fears that "at the practical and policy level . . . market incentives seem more determinative of the direction of research than do global norms of social accountability," in spite of the announcements of such statements. Cahill rightly observes that "transnational institutions and networks are exerting pressure on and limiting the influence of market incentives on genetic research," as already noted regarding women's equality, human rights, and the environment. While discussing Catholic social teaching about upholding the common good, distributive justice, and the like, she seems a little too optimistic about its influence on genetic research. In the societies where Catholic teaching and faith are more pronounced, it seems that there too self-interest and the interest of others clash in the same way and probably in more or less the same degree. What is needed is equity in the distribution of power (knowledge, skill, wealth, etc.). Without this equity it is not possible to control the market incentives of the corporate and multinational bodies with regard to genetics in particular.

This book, *Genetics, Theology, and Ethics: An Interdisciplinary Conversation,* gives us a clear insight into the relation between genetics, theology, and ethics. After reading the manuscript I personally have learned many aspects of the ethics of genetics in relation to theology. The book will benefit many in understanding in depth genome research and its consequences. I trust that the questions I have raised from the perspective of the developing world will help broaden the horizon of the discussion.

Notes

1. *Islamic Manual of Family Planning*, draft presentation in May 1996 in Cairo.
2. *Jugantar* (April 21, 2004): 17.
3. H. Begum, *Women in the Developing World* (New Delhi: Sterling, 1999); and eadem, *Ethics in Social Practice* (Dhaka: APPL, 2001).

Index